Seed Science and Technology

Editor
Gurnam Singh

2008
Gene-Tech Books
New Delhi - 110 002

ISBN10 81-89729-90-X
ISBN13 978-81-89729-90-5

Published by : **GENE-TECH BOOKS**
4762-63/23, Ansari Road, Darya Ganj,
NEW DELHI - 110 002
Phone: 41562849
e-mail: genetechbooks@yahoo.co.in

Printed at : **Tarun Offset Printers**
Delhi

PRINTED IN INDIA

Preface

Advancements in the field of plant science in the past decade has led to the birth of a whole new biological branch - seed science and technology. With radically evolving technologies in agriculture triggering quality control at each stage of the agricultural process, seed science and technology has gained immense prominence as a crucial factor in the quality control cycle.

This book has been designed to introduce to the reader the scientific principles underlying seed science and technology, by setting them against commercial (e.g., seed supply, weed control) as well as non commercial (e.g., habitat restoration) contexts. It seeks to make a broad understanding of the various pieces that make up the jigsaw puzzle of this science, and probes into the potential and the future of this industry.

This book begins the exploration and introduction of seed science by giving an overview of the basic elements like viability, vigour, quality and dormancy. The commercial and economic relevance of this industry is given an in-depth insight into. Apart from that, the different aspects of seed response to the environment, such as effects of pre- and post harvest environment on seed and consequences of subsequent tests, have been thoroughly probed. An awareness of the principles underlying seed quality testing for vigour and viability has been created, besides giving a panoramic view of the industry in its whole, and its increasing scope. It is hoped that, with a compendious presentation of the subject, readers associated with agricultural science would find this book highly useful.

Gurnam Singh

Contents

1

Seed Biology

Since ancient times, seeds have been symbolic in the religious creeds of human groups. In ancient Egypt, for example, seeds symbolised the mysteries of death, resurrection, and life after death. Mummies were buried with a vessel of viable seeds to fill the food requirements in a new life. Seed symbolism continues to be a part of religious rituals and ceremonies practiced by numerous people.

Gathering, storing, selecting, and sowing seeds capable of producing good crops became an art and an obligation: human food security was dependent on seeds. Throughout the centuries, seeds have been a means to reproduce species and a source of nutrients upon which all developing civilisations have been built. Seeds have ultimately produced fibres, carbo-hydrates, proteins, essential oils, insecticides, beverages, alkaloids, phytoestrogens, animal fodder, and other less important products.

Forest Tree Seeds

The forest and forest trees are not a boundless resource, and as we realise the value of treating the forest as a crop. The exhaustive, indiscriminate, and haphazard use of forest resources gradually limits the supply of traditional woods and other forest products important in national and international markets. Resource reduction also hastens the loss of a multiple use resource that beneficially affects soils, hydric regimes, biodiversity, landscapes, and microclimates. Additionally, the forest mass fixes CO_2, diminishes atmospheric contamination and deterioration, and contributes to sustaining human survival.

Today, increasing the existing knowledge about forest dynamics, trees as basic and essential units, and seeds and seedlings as indisputable factors of regeneration has become important. Challenging issues include generating new knowledge; developing effective techniques that promote conservation, management, and rational use of remaining forests; and implementing successful reforestation programmes. Most tropical tree pioneers, that is those typical of early forest succession, have small orthodox and dormant seeds. Easily dispersed, these seeds may form seed banks on the forest floor and may be photoblastic or thermoblastic.

Germination of most of these seeds is stimulated by environmental alterations or disturbances, such as natural or artificial clearings in the canopy, on different spatial or chronological scales. Changes in the rates of red/far-red light or temperature fluctuations produced by vegetation removal may increase the germination rate. In contrast, a large number of climax species have big recalcitrant or intermediate seeds. Recalcitrant seeds germinate quickly, an advantage in terms of avoiding insect predation and preventing lipid degradation from diverse microorganism action. A recalcitrant seed lacks maturation drying and its development is uninterrupted; however, the speed of embryo development (growth and differentiation) varies greatly among species.

Diaspore permanence on the forest litter or substrate before root protrusion may vary from days to months. For example, root protrusion in *Dipteryx panamensis* (Pittier) Record & Mell and *Pentaclethra macroloba* (Willd.) Kuntze occurs 8 to 10 days after sowing; in *Virola koschnyi* Warb., 11 to 14 days; in *Carapa* guianensis Aubl., 14 to 16; in Calophyllum brasiliense Cambess., 15 to 18; in *Lecythis ampla* Miers, 45 to 48; in *Ocotea* austinii C.K. Allen, 50 to 55; and in Minquartia guianensis Aubl., 150 to 160 (5 months or more). While pioneer species such as *Cecropia*, *Goethalsia meiantha* (Donn. Sm.) Burret, Hampea platanifolia, Heliocarpus appendiculatus, and H. americanus form seed banks on the forest floor, the species with recalcitrant seeds form seedling banks. The seedlings that remain until advanced stages of forest regeneration are, in many species, more important than the seed banks as a source of regeneration.

Regeneration includes dormancy types and germination times and types that are all obviously related to the morphology, physiology, and ecology of seeds. Consequently, the study of these aspects is becoming increasingly important to understanding seed behaviour and predicting the regeneration capacity of the trees and the forest. The completion of a life cycle or the regeneration of a plant community depends on the production of physiologically independent individuals, which happens through seed or vegetative reproduction. Establishment of the species reproduced by seeds is dominant in communities in early succession after strong disturbances, and in forests of intermittent regeneration, throughout a vegetation mosaic.

The genotype of a tree seed is the result of the evolutionary forces operating on the species for centuries. It is adapted to the present environmental conditions but not to those of the future, especially when environmental conditions are being drastically modified by accelerated global change. Even if scientific and technological knowledge could be increased, the forests of the past and those of today cannot be replicated tomorrow. The natural processes that gave rise to the species, communities, and ecosystems we want to preserve have been built over a long period of time and in large spaces. To preserve and reproduce tree species by seed, we must know the limits imposed on the ecophysiology of seeds by nature through the evolutionary process. The genetic combinations able to survive through environmental changes have intrinsic value and must be carefully selected.

Seed germination is also influenced by the environmental conditions during seed development and maturation while doron the parent tree. Day length, temperature, parental photothermic environment, light quality, and elevation are factors that significantly influence germination capacity. Additional factors include the inflorescence position on the parent tree, seed position in the fruit or infrutescence, and parent tree age during floral induction. These factors, plus others, explain the strong variation found in seed parameters (weight, colour, water content, germinability) among seed groups and among seeds in the same group.

To most people, the concept of "seed" is deceptively simple. It is compared to a pill, isolated from the environmental impact, replicated many times, and capable of producing a plant. However, a seed is attached to a long and complex evolutionary and physiological history, and it is also conditioned to the variations of a long and complex future. The study of seeds has many facets and should not be limited to collection, storage, and sowing.

Reproductive Biology

Reproductive Structures in Angiosperms

The flower is the reproductive structure in angiosperms. The classic flower has a pedicel, floral receptacle, calyx (group of sepals), corolla (group of petals), androecium (ensemble of stamens), and gynoecium (ensemble of carpels) (fig. 1). The androecium and the gynoecium are the fertile organs; the remaining organs are sterile (vegetative structures). Most flowers have a specific numeric arrangement of organs. In monocotyledons, the flowers generally have three pieces: three sepals, three petals, and usually three stamens or a multiple of three stamens; these flowers are called trimerous. The dicotyledons have flowers built in an array of four or five pieces and are called tetramerous or pentamerous. The flower plan is more evident in petals and sepals, and to some degree in the stamens; the plan is not applied to carpels. In the flower, the sepals alternate with the petals and the petals frequently alternate with the stamens. Groups of inner stamens may alternate with groups of outer stamens; sometimes the carpels are opposite to the sepals.

When all of the typical organs—petals, sepals, stamens and carpels—are present in a flower, it is complete; if one or more organs are missing, the flower is incomplete. A nude flower lacks a perianth; a flower is apetalous if it lacks a corolla (e.g., Roupala montana, Macadamia integrifolia, Oreomunnea pterocarpa, Sterculia apetala, Alfaroa costaricensis, Hyeronima alchorneoides). The terms regular and irregular are commonly used in reference to flowers. These terms generally involve the perianth (corolla + calyx); the reproductive organs are occasionally included. A regular, or actinomorphic, flower (e.g., *Coccoloba* tuerckheimii,

Terminalia amazonia (J.F. Gmel.) Exell, Ocotea austinii, Magnolia poasana, Gustavia hexapetala) has perianth pieces of similar shape and size.

The flower can be divided into equal halves by a vertical plane in several directions. This flower has radial symmetry (polysymmetric) (fig. 1). Bisymmetric flowers have two planes of symmetry (bilateral flowers; typical in Brassicaceae). The irregular, or zygomorphic, flowers are doron siventral or monosymmetric (e.g., *Andira, Diphysa americana* (Mill.) M. Sousa, Dipteryx panamensis, Erythrina, Gliricidia, Hymenolobium mesoamericanum, Theobroma cacao, Bertholletia, Cariniana, Couratari, Lecythis, Eschweilera, Qualea paraensis. The irregularity usually involves the petals, but the sepals and the perianth as a whole could be included. There are also asymmetric flowers. This asymmetry must be produced by organ reduction, multiplication, or modification (e.g., *Vochysia*, and typical in *Canna* and *Maranta*).

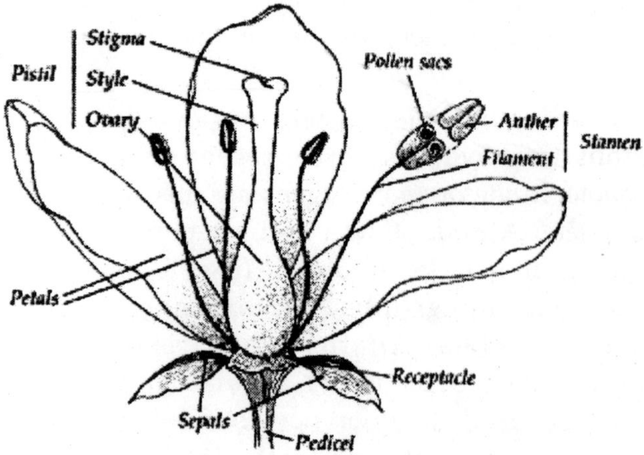

Figure 1. Perfect flower

Floral reduction can occur in one or several organs simultaneously. By reduction, organs may change their shape, structure, and function. The transformation of petals and stamens in glands, or of stamens in staminoids or petals, is the most common change. Stamen reduction may occur in different degrees, from anther abortion to organ disappearance. Carpel reduction usually refers to reduction in size and ovule number; for example, the achene is a fruit with only one ovule. Ovule reduction and loss is usually gradual; it may take place in acropetal direction, retaining only the distal ovule(s), or proceed in basipetal direction; in this case only the basal (proximal) ovule(s) remain. The receptacle resembles a shoot in ontogeny and structure. It has nodes, internodes, and appendages; the nodes are crowded due to internode reduction or suppression. The fertile and sterile floral organs arise at the receptacle nodes and are grouped in spirals or verticils.

The calyx and corolla are vegetative structures and collectively constitute the perianth. They are easily distinguished in many flowers (e.g., Xylopia sericophylla, Cananga odorata, Annona, Drimys granadensis, Guajacum sanctum L.); in other flowers the perianth parts are indistinguishable and are named *tepals* (e.g., *Cecropia*, staminate flowers; *Allocasuarina*, staminate flowers; *Hernandia didymantha, Magnolia poasana, Gyrocarpus*). In many flowers, the sepals and petals are modified leaves.

In many flowers, the calyx and corolla may be reduced, vestigial, or missing. Reduced sepals may be toothed, scaly, bristled, or ribbed. Petals are usually laminar and bigger than sepals. They vary widely in size and shape. Flowers with large and elaborate petals can be found in families such as Fabaceae-Caesalpinioideae (*Cassia grandis* L.f.), Hippocastanaceae (Billia), Lecythidaceae (Lecythis, Eschweilera, Grias cauliflora),

Fabaceae-Papilionoideae (Dipteryx oleifera, D. panamensis, Hymenolobium mesoamericanum, Hymenaea courbaril L.), and Vochysiaceae (*Vochysia, Qualea paraensis*).

Those with small, reduced, vestigial, or no petals are in families such as Betulaceae (*Alnus acuminata* Kunth in H.B.K.), Casuarinaceae (*Allocasuarina*), Euphorbiaceae (*Hyeronima*), Fagaceae (*Quer-cus*), Juglandaceae (*Alfaroa costaricensis* Standl.), Moraceae (*Brosimum, Maclura tinctoria* [syn. *Chlorophora tinctoria*], *Castilla elastica, Naucleopsis naga*), Myristicaceae (*Virola, Myristica fragrans, Otoba novogranatensis* Moldenke), Salicaceae (*Salix*), and Ticodendraceae (*Ticodendron incognitum*). When reduced, the petals may form scales, bristles, or glands of varied size and shape. The position of the calyx and corolla pieces in the floral bud (aestivation) sometimes persists in the mature flower. There are several aestivation types; the following classification follows that of Radford and others:

- *Alternate:* Two rows or series of organs or structures where the margins of the inner group are overlapped by a margin from each adjacent outer piece.
- *Cochleate:* A hollow or helmet-shaped structure encloses or surrounds the other pieces.
- *Contorted or regular:* Several structures or organs in a whorl or close spiral have one margin covering the margin of the adjacent structure.
- *Convolute:* A leaf or perianth piece partly rolled in another, commonly twisted distally.
- *Imbricate or irregular:* Piece margins overlap. Induplicate. Piece margins fold inward, touching the margins of the adjacent structures.

— *Quincuncial:* Five pieces: two are interior and two exterior; the fifth piece has one margin covering the interior pieces while the other margin is covered or surrounded by one margin of the exterior pieces.

— *Valvate:* Margins of the adjacent structures touch at edges without overlapping.

— *Vexillate:* Large piece surrounds other minor pieces, paralleloriented with respect to the outer piece semilimbs.

Perianth aestivation has taxonomic value; for example, the corolla aestivation is a feature distinguishing Caesalpinioideae from Papilionoideae.

The androecium is the ensemble of stamens. The typical stamen has the anther and the filament, which connects the anther to the floral axis or another organ. The anther comprises four elongated microsporangia (pollen sacs), where pollen develops. The androecium is outside the gynoecium and inside the corolla; the stamens are sometimes attached to or inserted upon the petals, the corolla (epipetalous), or the gynoecium (gynandrous). The stamens may be united through coherent or connate filaments in a single group (monadelphous), in two groups (diadelphous), or more (polydelphous) or by union of their anthers into a ring (syngenesious).

The attachment of the anther to the filament (fig. 2) is innate when the microsporangia are terminal and the filament is continuous between the microsporangia. It is adnate when the microsporangia are elongated and attached lengthwise to the filament. When the anther is attached by its middle, with both ends free, it is versatile. Dorsifixed anthers are attached dorsally and medially to the filament tip, while basifixed anthers are attached distally to the filament. In some species the microsporangia are in a transverse position. The number of stamen in a flower ranges from one to many, and from an indefinite number to a defined one.

The arrangement on the receptacle could be spiral (helical), whorled (verticillated), or fasciculate (clustered); the fascicles are commonly in whorls. The spirally arranged stamens represent a primitive condition from which the whorled and fasciculate types developed. The fascicles have various numbers of stamens and are formed by aggregation and conation of individual stamens to different degrees. When the stamens are in one whorl the androecium is haplostemonous. When the stamens are distributed in two whorls and the members of the outer verticil alternate with the petals, the androecium is diplostemonous; if the members of the outer whorl are opposite the petals, the androecium is obdiplostemonous. An androecium with more than two stamen whorls is polystemonous.

The stamens may vary in size and shape, but before dehiscence all anthers have two pairs of microsporangia in the anther lobules. The microsporangia are separated by a zone of sterile tissue called intervenal connective. In some species only one microsporangia develops; however, many bisporangiate anthers are truly tetrasporangiate. At maturity the two contiguous sporangia of each side flow together because the partition separating them breaks down.

All stamen parts can be modified in the evolutionary process. The filament length is variable and may be missing. The microsporangia may be reduced and sterile, and the stamens become staminodia. The zygomorphic flowers have only modified stamens or a reduced number of them. The gynoecium has a central position and may have one to several free or connate carpels. Each free carpel or gynoecium formed by several carpels has a stigma (structure receiving the pollen), a style (narrow portion below the stigma), and an ovary (wide base containing one or more ovules). Some gynoecia lack styles (e.g., *Clusia*, Clusiaceae), but ovaries and stigmas are always present. In the gynoecium, the ovules develop from the placenta. When the gynoecium has individual carpels, the flower is apocarpous (e.g., *Magnolia*); if carpels are connate, the flower is syncarpous (e.g., *Oreomunnea*).

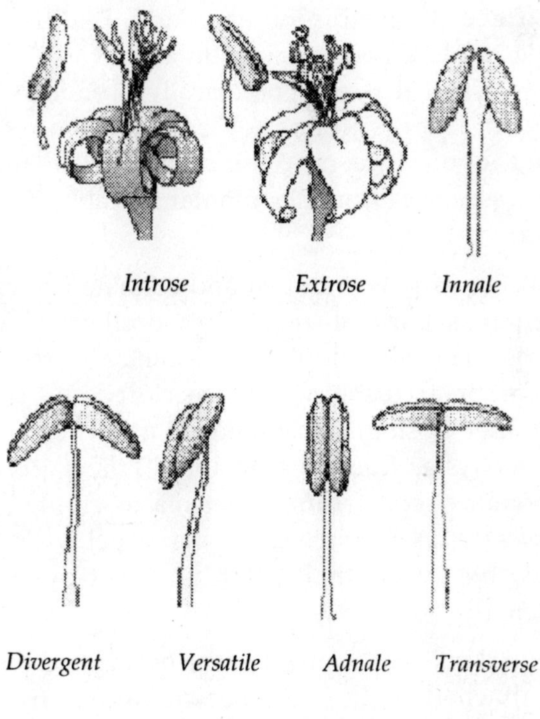

Figure 2. Anther Attachment.

Ovaries have at least two zones: the wall and the locule(s). Multilocular ovaries have a third zone: the septa or partitions. Each carpel has two placentas on the adaxial surface near the margins. When reduction of the carpelar margins is accentuated, the contiguous placentas join and sometimes fuse.

Placentation is the type of ovule distribution in the carpels forming the gynoecium. Carpels are located in the ovary. The primitive, free, laminar carpel has submarginal placentation because ovules develop in the placenta between the dorsal and ventral bundles. The closed carpel usually has laminar placentation such as in Mimosoideae, Caesalpinioideae, and Papilionoideae. When the ovule develops close to the ovary base it is basal even when it is morphologically pseudobasal (e.g., Calophyllum brasiliense, Virola koschnyi). The suspended ovules are near the distal end of the ovary (e.g., *Allocasuarina*, Roupala montana Aubl., Terminalia amazonia).

When the carpels are open and connate (syncarpous), placentation is parietal (e.g., Bixa orellana, Escallonia myrtilloides, Hasseltia floribunda, Homalium racemosum, Casearia arborea); if the carpels are closed and connation occurred after closing, placentation is axilar (e.g., Lecythis ampla, Vochysia, Luehea seemannii Triana & Planch., Pseudobombax septenatum, Swietenia macrophylla King, *Cedrela odorata* L.). In some cases, the carpel walls or septa dividing the ovary in locules are destroyed during development.

A central column appears in the ovary, and the ovules appear attached to it. This placentation is free-central (e.g., *Theobroma cacao*). If the carpels forming a syncarpous gynoecium were open before connation, usually only a central locule forms. Generally, a carpel has three vascular bundles: a dorsal (median) vein and two ventral (lateral, marginal) veins. Many carpels have additional veins while others have lost veins by reduction and have fewer than the basic three. When the gynoecium is in the distal end of the floral axis, the ovary is superior and the flower hypogynous (Fig. 3).

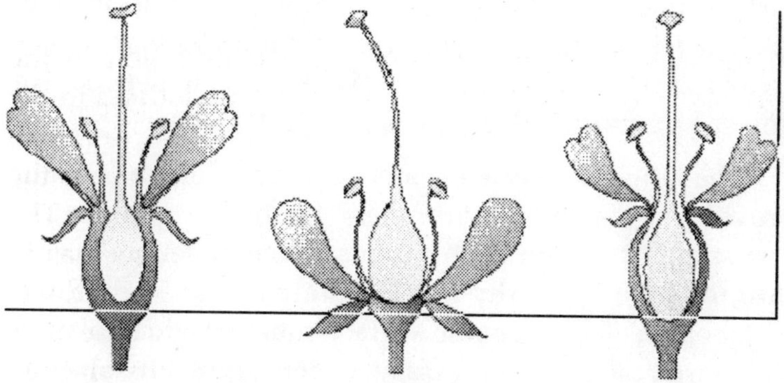

Figure 3. Ovary position and flower types.

When the gynoecium occupies the lowest level in the floral axis and its wall is adnate to the floral tube or hypanthium to which the other floral organs (perianth and androecium) are attached, the ovary is inferior and the flower is epigynous. Some flowers have the perianth and androecium attached to the floral tube or hypanthium surrounding the ovary without fusing to it. These flowers are perigynous and the ovaries are semi-inferior. They are transitional forms to epigyny. More complicated classifications include several subtypes.

The style is the narrow carpel zone located between the ovary and the stigma. In the syncarpic gynoecia that bear a single style, tissues of the carpels integrating the gynoecium form the style. The carpels may be partially fused distally, where the style is a single structure at the base and a compound one at the top. The style may have stylar branches (stylodes); usually, one stylode per carpel (e.g., *Hibiscus*), and the styles or stylodes can be solid or have a central canal (hollow). The stigma is the receptive surface that captures the pollen grains. It is formed by specialised surface cells that connect to the stylar tissues. Style and stigma morphology is diverse; it depends on flower structure and pollination type (Fig. 3).

The stigma captures and accepts the pollen and provides a suitable environment for pollen germination. The stigma can provide nutrients to the pollen and orient pollen tube growth; it also helps regulate flower metabolism. The stigmatic surface must provide balanced osmolarity and the necessary water supply to promote pollen germination. Frequently, changing environmental conditions delay pollination, and the style and stigma must maintain the vitality to continue functioning.

The pollen wall and the stigmatic surface are involved in the cellular reception and recognition of signals or stimuli. Some of the macromolecules involved in this mechanism are allergic substances, stigmatic antigens, carbohydrates, proteins, gycoproteins, and glycosphingolipids. Stigmas are divided into two categories: wet stigmas, with a copious fluid secretion; and dry stigmas, with a slight fluid secretion.

The dry stigmas have receptive cells dispersed in multiseriate stylodes or concentrated in ridges, zones, or heads. Those with specialised reception areas may be separated into papillate and nonpapillate stigmas. Wet stigmas have a receptive surface with small or medium-sized papillae. The binucleate pollen grains seem correlated to wet and dry stigmas, while trinucleate pollen grains are associated more with dry stigmas.

Sexual Expression in the Flower, the Inflorescence, and the Sporophyte

The morphology and physiology of the flower in tropical trees are varied and are associated with pollination and the sexual systems of the species. Plants can be either self- or cross-pollinated; however, sex separation in space or time, as well as selfsterility (homogamy) are frequent mechanisms used to favour cross-pollination. In the individual flower the separation of sexes in space has three modalities:

Perfect, bisexual, or hermaphrodite: Flowers with pistillate (gynoecium) and staminate (androecium) organs. Examples include: Persea americana Mill., Dipteryx panamensis, D. oleifera, Samanea saman, Stryphnodendron microstachyum, Vochysia guatemalensis Donn. Sm., Pentaclethra macroloba, Terminalia amazonia, Lecythis ampla, and Sideroxylon persimile.

*Unisexual:*Single-sex flowers with androecium or gynoecium. Examples include Brosimum alicastrum Sw., Diospyros nicaraguensis, Hura crepitans L., Hyeronima alchorneoides Allemao., and *Virola koschnyi.* In species such as *Carapa guianensis* Aubl., the flowers are unisexual (male or female) but maintain vestiges of the other sex. There are two subtypes of unisexual flowers:

(1) *Staminate or male*—Flowers with an androecium but no gynoecium. For example: *Batocarpus, Casuarina, Hyeronima, Otoba novogranatensis, and Virola.*

(2) *Pistillate, carpellate, or female*—Flowers with a gynoecium but no androecium. For example: *Batocarpus, Casuarina, Hyeronima, Otoba novogranatensis, and Virola.*

Neutral or agamic: Sterile flowers with abortive or no sexual organs.

In the inflorescence, the spatial separation of sexes may have numerous variations and combinations. In the individual tree, the sexual expression must be as follows:

Hermaphrodite or monoclinous: Tree with only hermaphrodite flowers (bisexual). For example: *Pentaclethra macroloba, Stryphnodendron microstachyum,* and *Terminalia amazonia.*

Monoecious or diclinous: Tree with both staminate and pistillate flowers. For example: *Artocarpus, Carapa guianensis, Hernandia didymantha, Poulsenia armata,* and *Ricinus communis.*

Dioecious: Tree with only one type of unisexual (imperfect) flowers. For example: *Virola, Hyeronima, and Batocarpus.* There are two subtypes:

(1) *Gynoecious,, pistillate, or female*—Trees with only pistillate flowers. For example: *Guarea rhopalocarpa, Hampea, Virola,* and *Hyeronima.*

(2) *Androecious, staminate, or male*—Trees with only staminate flowers. For example: *Ilex, Myristica fragrans, Siparuna grandiflora, Virola koschnyi,* and *Hyeronimu alchorneoides.*

Androdioecious: Some trees with hermaphrodite flowers and others with staminate or male flowers.

Andromonoecious: Tree with both hermaphrodite and staminate flowers. For example: *Calophyllum brasiliense, Parkia bicolor,* and *Schefflera.*

Gynodioecious: Some trees with hermaphrodite flowers and others with pistillate or female flowers. For example: *Ocotea* tenera.

Gynomonoecious: Tree with both hermaphrodite and pistillate flowers.

Polygamous or trimonoecious: Tree with hermaphrodite, pistillate, and staminate flowers. For example: *Aesculus hippocastanum and Gyrocarpus jatrophifolius.*

Polygamo-dioecious: Staminate, pistillate, and hermaphrodite trees in the same species. For example: *Coccoloba uvifera (L.) L.* In many plants, a mechanism under genetic control prevents the ovules being fertilised by pollen from the same tree. In these species, the pollen may reach the stigma of its own flower but fertilisation does not occur. In dioecious species, cross-pollination is the only alternative; in monoecious species, the separation of sexes into different flowers promotes but does not guarantee cross-pollination.

The separation of sexes by time prevents self-pollination in hermaphrodite flowers; the stamens and stigmas mature at different times, diminishing the possibility of self-pollination. The separation of sexes by time is known as dichogamy. There are two modalities of dichogamy:

Protandry (protandrous flowers). Stamens mature first and pollen release occurs before stigmas are receptive. Protogyny. Stigmas mature first and are receptive and functional before anther dehiscence and pollen release.

Protandry is common in legumes (Fabaceae) and Malvaceae; protogyny is frequent in many families including Moraceae (*Ficus*), Rosaceae, Annonaceae (some *Annona*), Chrysobalanaceae, Combretaceae, and Lauraceae. In Lauraceae, the flowers are protogynous (proterogynous) and the floral mechanism in the family is the synchronised dichogamy. In the family, the species have two types of flowering in different individuals. In some, the flowers have receptive stigmas (opening of the female phase) in the early morning, ceasing activity at noon (closing of the female phase). In the afternoon of the same day or the following day, the stamens release pollen (opening of the male phase), ceasing activity at dusk (closing of the male phase). In other trees, the flowers have receptive stigmas in the afternoon and cease activity at dusk.

The following morning, the stamens release pollen. This mechanism preventing self-pollination and cross-pollination becomes obligatory. If the rhythm is disturbed by changing climatic conditions, the flower could be pollinated and fertilised by pollen coming from another flower in the same tree (geitonogamy), but the fruits undergo an early abscission. A postzygotic mechanism of incompatibility has been suspected. Dichogamy has also been documented in Annonaceae [*Annona squamosa*, *A. cherimolia*, *A. reticulata*, *Cananga odorata*, *Rollinia membranacea* (syn. *R. jimenezii*), Betulaceae (*Alnus acuminata*) and Combretaceae (*Terminalia amazonia*)]. *Terminalia amazonia* has protogynous flowers. It is believed that the flowers are pollinated by pollen coming from the flowers of other trees (xenogamy); however, sporadic geitonogamy appears to occur. In this species, a postzygotic mechanism of incompatibility has also been suggested.

In numerous species, the dichogamy is partial and the second sexual phase begins before the first sexual phase is over. The flowers are functionally unisexual in the initial phase and bisexual (hermaphrodite) in their terminal phase. For example, the flowers of *Symphonia globulifera* L.f. are protandrous, with a bisexual phase in which the stigma is receptive and the anthers release pollen.

The homogamous (self-sterile) hermaphrodite flowers release and accept pollen at the same time; nevertheless, they are not always autogamous. Both morphological (heteromorphic autoincompatibility) and physiological (homomorphic autoincompatibility) mechanisms may prevent self-pollination, self-fertilisation, or seed development, thus promoting self-sterility. Heteromorphic self-incompatibility is characterised by intraspecific differences in floral structure, reinforcing cross-pollination and cross-fertilisation. Floral heterostyly is the most typical mechanism. In this case, two or three floral morphs exist. A distylous species has two floral morphs; in some flowers the styles are long, in others they are short.

In the flowers with long styles, the stamens are found below the stigma level while in those with short styles, the stamens are above the stigma level. Other features are usually associated with style length. Morphs with long styles and short stamens have small pollen grains and long stigmatic papillae; morphs with short styles and long stamens have big pollen grains and small stigmatic papillae. These features favour crosspollination between the two floral morphs. However, some species have a diallelic mechanism of incompatibility, lacking stylar and staminal polymorphism, and others have a polymorphism that is reduced to the style.

The homostyly of neotropical laurel [*Cordia alliodora*, Boraginaceae] flowers has been described repeatedly; however, the species has a kind of heteromorphy restricted to stylar variation, and the small flower size may hinder differentiation between style and stigma. Additionally, two tree groups exist in which intercrossing is possible but intracrossing is not. The low variation in stylar length must be associated with a diallelic incompatibility system in the sporophyte.

Heterostyly is a type of genetically controlled polymorphism shared by at least 25 families of angiosperms, e.g., *Byrsocarpus, Rudgea, Psychotria, Turnera ulmifolia, Erythroxylum coca,* and *Cratoxylum.* In distylous species, the polymorphism acts in conjunction with a diallelic incompatibility system to prevent self- and intramorphic fertilisation. Homomorphic auto-incompatibility can be divided into two subtypes:

Gametophytic: Fertilisation is inhibited as a result of genetic action in the pollen grain (male gametophyte) when the pollen tube protrudes through the stigmatic-stylar tissues *(e.g., Tectona grandis L.f.).*

Sporophytic: Inhibition of pollen germination or pollen tube growth is imposed by genetic action of the sporophytic tissues such as the stigma, the style, and sometimes the embryo sac; e.g., *Dalbergia retusa* Hemsl., *Dalbergia miscolobium, Dipteryx panamensis, Myrospermum frutescens* and probably *Terminalia amazonia.*

The incompatibility systems share a common factor: both maintain a high degree of genetic heterozygocity in the species population. With a few exceptions, the incompatibility system operates at the stigma or style level. Because both structures have diploid tissues and a sporophyte genome (parent tree), the incompatibility reaction is controlled by the genome of the parent tree. The male gametophyte (pollen grain) has in part the genome of the sporophyte in which it develops and some sporophytic components.

Thus, the incompatibility reaction of the pollen can be controlled by the gametophyte itself or by the sporophytic components carried with it. In the gametophytic incompatibility the haploid genome of the pollen and the diploid genome of the stigma/style genetically control the pollen-stigma/style interaction. Incompatibility that occurs in the embryo sac is haploid-

haploid. The sporophytic incompatibility is determined by the genome reaction of the paternal sporophyte from which the pollen grain originated.

In some species, homogamous flowers can be cleistogamous. In these flowers, pollen release and stigma receptivity are synchronised before floral perianth opening. Species with this type of flower include *Annona muricata, Cyathocalyx, Dasymaschalon, Goniothalamus* sp. in Annonaceae, *Lithospermum caroliniense,* and *Coffea arabica* (Rubiaceae).

If the maturation and functioning of the unisexual flowers (staminate and pistillate) in a monoecious tree are not synchronised, cross-pollination is obligatory. In dioecious species, cross-pollination and cross-fertilisation are obligatory. The forests in the lowlands of the Atlantic coast in Costa Rica show dominance of species with hermaphrodite flowers in the canopy and subcanopy; dioecious species have a lower representation and monoecious species are scarce.

Inflorescence Types

Because the traditional nomenclature used to describe inflorescences, which shows morphological inconsistencies, is widely used, descriptions and definitions follow. The stem bearing the inflorescence is the peduncle, while that bearing the flower is the pedicel. The flowers develop in the bract axil; if the inflorescence lacks bracts it is ecbracteate. The groups or whorls of bracts form an involucre; secondary involucres, as in the compound umbels, are involucels of bractlets. When a single conspicuous bract subtends a floral group (usually a fleshy spike or spadix), it is called a spathe.

The most common types of traditional inflorescences are the following:

Dichasium: Peduncle with a terminal flower and a pair of branches, each producing a single flower. The central flower matures first. The simple dichasium can be repeated as a unit several times giving rise to a compound dichasium. The dichasium has been named cyme, but the term is used indiscriminately and its use is inappropriate.

Monochasium: Peduncle with a terminal flower and a basal lateral branch bearing a single flower. The simple monochasium can be replicated several times giving rise to a compound monochasium, which may form the following subtypes:

(1) *Bostryx (helical monochasium)*—Flower distribution spirally coiled around the main axis.

(2) *Cincinnus*—Flowering alternately to one side and then the other along one side of the axis, the whole inflorescence often coiling downward.

(3) *Rhipidium*—A bostryx is extended in one plane, and the inflorescence is sometimes fan-shaped.

(4) *Drepanium*—A cincinnus bears all the branches on one side. The inflorescence extends in one plane and coils downward to one side.

Panicle: Inflorescence with a central axis giving rise to lateral branches that are themselves branched. The minor ramifications can be dichasial. A crowded, nearly cylindrical panicle is called a thyrse.

Raceme: Inflorescence with a central axis, along which simple pedicels of similar length are located. The blooming usually proceeds upwards, although it may be irregular. Spike. Elongated inflorescence with an unbranched central axis with sessile or subsessile flowers. The blooming usually proceeds upwards. The spikelet or locusta is the basic unit of the inflorescence in grasses and sedges.

Corymb: Flat-topped inflorescence with a central axis giving rise to branches of unequal length. The lateral branches may branch further or be simple pedicels. Blooming usually begins in the external flowers and proceeds centripetally, but can also be irregular.

Head (capitulum): A determinate or indeterminate, rounded or flat-topped cluster of flowers placed at the same level. The flowers can be sessile or subsessile and are placed on a wide receptacle. Blooming usually begins at the periphery and proceeds centripetally. Heads may be solitary or form diverse aggregates. Each bract of the involucre is a phyllary.

Umbel: Inflorescence with several branches arising from a common point, in the distal end of the peduncle. If each branch ends in a single terminal flower, the umbel is simple; if it ends in a secondary umbellate, the umbel is compound. Flowering usually begins in the peripheral flowers and proceeds centripetally. The umbel may have an involucre at the peduncle distal end. If the umbellates have similar whorls of bracts, they are involucels and each bract is a bractlet. The main branches of a compound umbel are rays, and the corresponding members of the umbellate are pedicels.

Ament (catkin): Spike, raceme, or dichasium of apetalous unisexual flowers. It can be long or short, erect or hanging. The flowers are usually very small and subtended by a scaly, small bract. Spadix. Spike with a thick, fleshy apex, surrounded by a large, usually colored bract (spathe). Frequently, the flowers are very small and unisexual.

Siconium (hypanthodium): Inflorescence with small flowers on the wall of a concave head.

Other inflorescence types include:

Glomerule: Inflorescence formed by an indeterminate cluster of sessile or subsessile flowers.

Verticillaster: A group of whorled dichasia at the nodes of an elongated axis.

Pseudanthium: Inflorescence with several flowers simulating a single flower but composed of more than a single axis bearing subsidiary flowers. The pseudanthium is subtended by an involucre: scapose, with a single flower or inflorescence on a leafless peduncle; or scape, usually arising from a basal rosette.

Reproductive Cycle

During the flowering period of angiosperms, the tree or sporophyte produces microspores forming microgametophytes or pollen grains, and megaspores producing megagametophytes or embryo sacs inside the ovules. In most angiosperms the entire reproductive cycle, from floral inception to fruit and seed maturation, occurs in one season; however, there are exceptions such as the red oak (*Quercus* subg. *Quercus*) and *Allocasuarina*.

Pollen

The embryonic anther has a mass of fundamental tissue surrounded by a protodermis. This layer forms the epidermis of the anther. In the four corners of the developing anther, two to four hypodermal cells are formed; each group divides periclinally forming a series of primary parietal cells and primary sporogenous cells (internal layer). In the mature anther the outer layer or epi-dermis covering the microsporangium wall may remain intact in families such as Lauraceae, Magnoliaceae, and Winteraceae. In others, such as Moraceae and Ulmaceae, the epidermal cells restrain or fall down during anther maturation.

In some species, the epidermis has trichomes; in others, it develops fibrillar bands and is called an endothecium. If the epidermis falls down, the underlying layer substitutes for it. This layer is commonly called the endothecium or fibrous layer because it has cell wall thickenings; these thickenings are absent in the stomium. Middle layers consist of one or several layers of tubular cells with thin walls beneath the endothecium. These layers are compressed or crushed during microspore formation. The inner layer of the microsporangium wall is the tapetum. It seems to accomplish three functions: microspore nutrition, exine formation, and synthesis of some materials contributing to pollen grain wall formation (tryphine and pollenkitt).

In angiosperms the tapetum can be glandular (secretory) or periplasmodial (amoeboid). In the first case, the cells remain intact and persist in situ; in the second, the cell walls break down and the protoplasts mingle. The protoplasmic mass forms a multinucleate periplasmodium, which becomes mixed with the pollen mother cells. The cells in the primary sporogenous layer can divide before microsporogenesis, or directly become the microsporocytes or pollen mother cells. These, through meiosis, produce haploid microspores (n) which later differentiate as pollen grains. The cells in the primary parietal layer form, through periclinal and anticlinal divisions, a variable number of layers concentrically placed, which differentiate and become the several layers of the mature sporangium wall.

Meiosis of the pollen mother cells in the tapetum is the time of the synthesis of substances such as starch, lipids, and tryphine. The tapetum is involved in numerous activities. It keeps a delicate balance with the differentiation of the sporogenous tissue, and any alteration in it induces the degradation of pollen. Some evidence indicates that, through phys-ical or physiological factors, the tapetum may induce the abortion of pollen. The endothecium or the vascular bundle of the filament may influence the abortion process. The carotenoids and carotenoid esters formed in the tapetal cells participate in the formation of the sporopollenin, a major component of the pollen wall.

The pollenkitt formed by lipids, glycolipids, glycoproteins, monosaccharides, flavonoids and carotenoids synthesized in the tapetum, gives adhesiveness, order, and color to the pollen grain. Its probable functions are insect attraction, pollen protection against ultraviolet radiation, and adhesion to insects' bodies; the last function contributes to pollen dispersal. The pollen grain is the structure containing the male gametes. It is formed by a vegetative cell (tube cell) and a generative cell immersed in the vegetative cell cytoplasm (bicellular pollen grain or microgametophyte); both cells are haploid. If the generative cell divides mitotically and gives rise to the spermatic cells during pollen maturation and before pollen tube development, the pollen grain is tricellular.

The nucleus of the vegetative cell controls the pollen tube's metabolism as well as its initial growth after germination. The spermatic cells have self-motility, are structurally different, and may have different patterns of cytoplasmic genetic transmission. The pollen grain covering is complex and is formed by the exine (external layer formed by sporopollenin and characterised by a specific ornamentation) and the intine (smooth inner layer made of polysaccharides). The layers are chemically, morphologically, and ontogenetically different. Cell wall formation follows different patterns in different species, as an adaptation to various factors including cellular environment, pollination method, and potential interactions with the different pollination vectors.

The correlation between exine structure and pollination method has been corroborated in diverse families pollinated by both wind and animal vectors. The transportation methods and the structures expediting pollen adhesion to the stigma are adaptations; for example, exine thickness and ornamentation, aperture type, and wall adhesives. Pollen-stigma recognition and the allergic reaction in humans (pollen wall proteins and glycoproteins) are also adaptations. Pollen germination and pollen tube growth throughout the style involve the action of specific enzymes and heterotrophic nutrition; these conditions represent a high degree of specialisation.

The pollen of most angiosperms is a free cellular structure (monad), but in some families there are compound grains. Compound grains can be arranged in tetrads (a microsporocyte product), polyads of grains, or complex masses (pollinia). The tetrads may have zero to four fertile grains mixed with a remnant of abortive grains. Pollen shape, a genetically determined feature, is influenced by three factors: microspore position in the tetrad, number and disposition of germinal apertures, and differential cellular expansion during microspore and pollen grain development.

Anther dehiscence is the mechanism of pollen release. In some families the pollen is released through a small opening usually placed at the distal end of the anther. This type of dehiscence is poricide (e.g., *Hyeronima alchorneoides*, *Conostegia*, *Miconia*, *Gustavia*). In most species, however, the anther dehisces longitudinally along the stomium (longitudinal dehiscence; e.g., *Vochysia, Alnus acuminata*) (fig. 4).

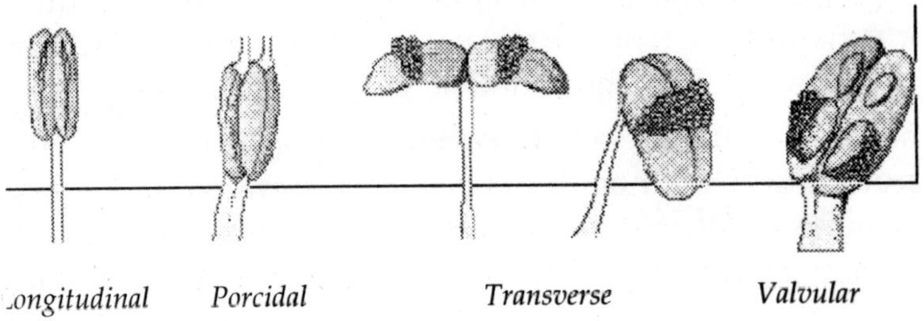

Figure 4. Anther dehiscence

If the anther is oriented outward and dehisces longitudinally to the exterior, it is called extrorse; if it is oriented inward and dehisces longitudinally to the interior, it is introrse (*Lacandonia*). In other species the anther opens transversely (transverse dehiscence; e.g., *Brosimum alicastrum*). Although the force opening the stomium seems to be generated in the endothecium, this issue is still a matter of discussion. The dehiscence is not always produced by changes in the fibrous layers; in many anthers, especially those of the poricide type, this layer is missing. Anthers with a complex shape are usually associated with special types of poricide dehiscence and have tubular projections in the pollen sac ends, through which the pollen is released.

The anthers with elongated, narrow openings may have a valve. This type of dehiscence is valvular and occurs in Hernandiaceae (*Hernandia stenura*), Lauraceae (*Ocotea, Nectandra*), and Monimiaceae (*Mollinedia, Siparuna*) (fig. 4). The valve opens at the base and projects outward and backward, carrying the pollen mass outside. The number and distribution of valves has taxonomic value in Lauraceae.

Ovule

The nucellus (megasporangium), integument(s), chalaza, and funiculus form the typical ovule. The nucellus occupies the central part of the ovule and is surrounded by the integuments. These leave a passage (micropyle) for pollen tube penetration. The integuments protect and nourish the nucellus. After fertilisation, they form the seedcoat, the raphe, and the chalaza. The ovules can be bitegmic, unitegmic or ategmic.

In some Annonaceae (Cananga odorata, Cleistopholis patens, Cyathocalyx sumatranus, Guamia mariannae, Mezzetia leptopoda, *Mezzettopsis creaghi*) a third integument, placed between the testa and the tegmen, has been proposed. The third integument in the Opuntieae (Cactaceae) is a funicular covering surrounding the seed. The funiculus is an extension, usually filamentous, that unites the ovule with the placenta. The funiculus has a vascular bundle transporting nutrients from the sporophyte to the megagametophyte.

Frequently, the ovule has special structures:

Hypostase: Usually cupuliform, formed by differentiation of nucellar and chalazal tissues. It is located on the chalazal vascular bundle and can be a cell mass or a discoid cell plate. It is found in the ovules of the Anacardiaceae, Bixaceae, Euphorbiaceae, and Lauraceae. It is considered a physical barrier to retain embryo growth and a bridge connecting the chalazal vascular bundle with the embryo sac, which facilitates nutrient transfer to the embryo sac. Other functions include producing enzymes and hormones and maintaining water balance in dormant seeds. Podium. A nucellar remnant, resistant to the absorbing activity of the embryo sac. This structure is located at the nucellar base and is cupuliform in shape. When it is located near the located near the micropylar end, it is epistase. After fertilisation, it generally precludes the micropyle.

Obturator: Formed by funicular, placental, or funicular-placental tissues. It is a protuberance covered by secretory trichomes or epidermal papillae, functioning as transfer cells. The obturator surface extends to the micropyle or enters through it. In general, it is continuous with the ovarian, stylar, and stigmatic tissues through which the pollen tube penetrates. After pollination, the obturator degenerates.

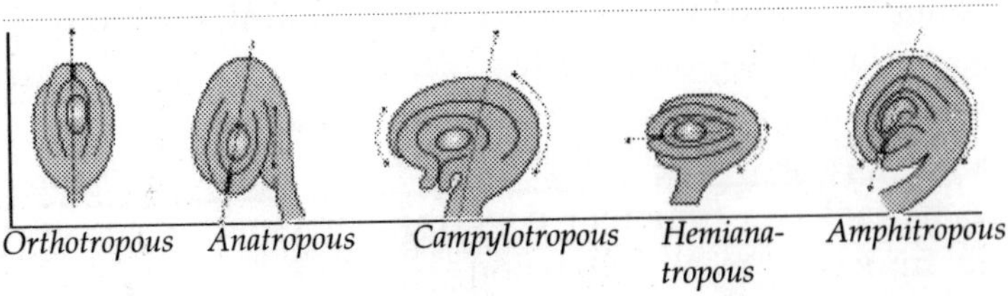

Figure 5. Ovule types

The ovules have different forms and degrees of curvature: athropous (orthotropous), anatropous, campylotropous, hemianatropous, or amphitropous (fig. 5). In general, the differential growth is in the funiculus-chalaza area. The integuments and the nucellus cause the differences in shape and curvature exhibited by the ovules and the mature seeds. In many cases, the curvature shown by an ovule during the floral anthesis is different and usually smaller than that shown by the mature seed. After fertilisation, the differential growth occurring in the different parts of the ovule may increase or diminish the ovule curvature. The seed curvature is correlated to embryo shape, micropyle position, and hilum length.

During ovule development, the archesporial cell differentiates in the nucellus. This cell originates, directly or indirectly, in the megasporocyte, meiocyte, or megaspore mother cell. The latter divides by meiosis and gives rise to four haploid megaspores. There are many variations in the megasporogenesis, but in most cases only the chalazal megaspore is functional. Mitotic divisions of the functional megaspore produce the monosporic megagametophyte (embryo sac). This type is common in the angiosperms. In other cases, no cell wall forms during the second mitotic division and a dyad is formed. The binucleate cell serves as a starting point for the formation of the embryo sac (bisporic). There are species in which no wall forms during the meiosis and a coenomegaspore with four nuclei is formed; it serves as the starting point for embryo sac formation (tetrasporic).

Approximately 60 to 70 percent of angiosperms have an embryo sac with eight nuclei and seven cells: the egg, two synergids, a central cell (binucleate), and three antipodals. The egg cell and the synergids are big cells located in the micropylar end of the embryo sac and forming a structure called the egg apparatus. The synergids are partially surrounded by a wall that is thick at the micropylar end; this thickening forms the filiform apparatus.

The central cell occupies the middle of the embryo sac and contains two polar nuclei or a fusion nucleus; the cell wall is partial and is restricted to the chalazal end. The cell membrane or plasmalemma has the typical invaginations of a transfer cell at the micropylar end. The antipodals are small cells with a varied structure, and may be ephemeral or persistent after fertilisation. They transport nutrients from the nucellus to the central cell and store nutrients for future endosperm. These cells may also serve a secretory function including secretion of growth regulators involved in endosperm development.

Flower Pollination and Fertilisation

Pollination Vectors

In angiosperms, pollination can be biotic or abiotic; most species participate with more than a single vector. For example, in *Mabea fistulifera* (Euphorbiaceae), pollination occurs through diurnal (passerine birds, bees) and nocturnal (bats and *Didelphis marsupialis*) pollinators. The adaptation to pollination mechanism is controlled by numerous evolutionary principles. Floral morphology is related to the most efficient pollinator, and evolutionary adaptation is parallel between the pollination vector and the structural, spectral, and olfactory specifications of the flower.

Biotic pollination must be carried out by different vectors such as insects (entomophily), birds (ornithophily), and bats (chiropterophily). Several mammals contribute to a minor degree, including prosimians, marsupials, rodents, giraffes, reptiles such as lizards, and invertebrates such as snails and slugs (malacophily). Biotic pollination provides mutual benefits: the vector obtains a reward (nectar, fragrances, oils, or pollen) and the plant disperses the pollen.

The abiotic agents are gravity, wind [air movements or air currents (anemophily)], and water [movements or currents (hydrophily)]. Pollination by gravity is unidimensional and, in general, results in species autogamy. Dispersal by wind is tridimensional; by water it can be bidimensional if it occurs on the surface or tridimensional if large water volumes or drops are involved. The different pollination vectors are associated with floral syndromes or groups of characteristics displayed by the diversity of flowers to attract and reward them.

The most common means of attraction are colors and fragrances. For example, most hymenopterans have similar vision systems of color, with three kinds of receptors (ultraviolet, blue, and green) and only some species have tetrachromic vision. Floral color is not restricted to the perianth; it may be found in bracts [*Bougainvillea* (Nyctaginaceae), *Euphorbia pulcherrima* (Euphorbiaceae)], stamens, and gynoecium. The colors can be produced by light refraction on physical structures and surfaces or by absorption of a defined light wavelength by the pigments.

The carotenoids, which are aliphatic compounds, produce colors varying from light yellow to dark red. The flavonoids (anthocyanins, flavones, flavonols, aurones) produce a color scope that ranges from light yellow in the flavones to purple and blue in the anthocyanins. Most color displays and chemical mechanisms use a limited number of anthocyanins (flavonoids) and include the anthocyanin color loss and the participation of a copigment. The most common anthocyanins are the magenta cyanidine, the purple delphinine, and the red pelargonin. The betalains are alkaloids producing a group of colors varying from red to light ivory. The color changes produced by age can be an indicator for vector visits, but the vector does not always discriminate on this basis.

Nectar production influences vector behaviour in the pollinating process. Nectar is composed of monosaccharides (glucose, fructose) and disaccharides (maltose, melobiose), amino acids, proteins (enzymes), lipids, alkaloids, phenols, antioxidants (ascorbic acid), organic acids, saponins, dextrins, and inorganic substances. The nectar can be directly ingested by the visitor (birds, bats, lepidopterans, dipterans) or carried to the nest to feed the larvae (hymenopterans). A strong correlation exists between the concentration of sugars and amino acids in the nectar and the type of pollinator.

In addition to pollinator reward, the nectar may have other functions. In some plants it serves as substrate for pollen grain germination; in others, inhibiting yeasts present in the nectar may inhibit pollen germination. The nectar also attracts insects to protect the plant against predators, a phenomenon present in the Neotropics. Flowers with elaiophores offer oils. The oils are formed in epithelial glands of the calyx (most New World Malpighiaceae) and released as an exudate exploited by bees such as *Centris* (e.g., *Byrsonima crassifolia* (L.) Kunth, *B. crispa,* some *Bunchosia*) and Krameriaceae. These oil-producing flowers are pollinated by oil-collecting bees. The plant may use mechanisms, such as inflorescence size, floral morphology, anthesis patterns, nectar production, and dichogamy, to restrict pollen remotion.

Pollinating insects can be coleopterans [beetles (cantharophily)], hymenopterans (symphytophily), wasps (vespophily), ants (phormicophily), bees (melittophily), flies (myophily), butterflies (psychophily), and moths (palaenophily). Some insects have mechanisms to answer the optical signals sent by the anthers, confirming that stamens and their mimetism are common signals for pollinating insects. Fragrances also attract insects; for example, the osmophores of the neotropical orchid *Stanhopea* and other genera, such as *Clowesia* and *Polycnis,* produce floral scents that attract euglossine bees. These scents consist of several compounds such as monoterpenes and benzenoids.

The production of heath-promoting aroma volatilisation to attract pollination vectors has been corroborated in numerous tree species. For example, the flowers of *Annona muricata* and other species in the genus have thermogenic respiration which, combined with strong fragrances, attract dynastid insects. The male euglossine bees (*Euglossa, Eulaema*) collect fragrances from flowers, mix them with the labial glandular secretions, and probably biochemically modify the fragrances to attract females for mating. The scent has a daily cycle, probably influenced by the thermal environment. These variations may influence insect distribution and mating systems. Some insects can change their preference for scents with the season. The entomophilous species usually have large, spherical pollen grains and a thick, ornamented exine which facilitates the adhesion to insects and stigmatic surfaces. Many pollen grains also have an adhesive cover or lipid seal (pollenkitt).

The coleoptera are a primitive group and cantharophily has been associated with the most primitive angiosperm families; nevertheless, in some families such as the Annonaceae, Calycanthaceae, Eupomatiaceae, and Magnoliaceae, the cantharophily is derived, secondary, and parallel to floral specialisation. In those groups, the flowers are open, have no structural restrictions to insect access, and produce an excess of organs and tissues that are eaten by the beetles. Only exceptional cases have flowers with nectar.

The Myristicaceae from southeast Asia (*Myristica, Horsfeldia, Knema, Gymnacranthera*) also seem to be pollinated by coleoptera even when they have a floral structure that strongly differs from that of the annonaceous flowers. The flowers are whitish or creamy, have a nocturnal scent, and lack nectaries or defined secretory structures. They are unisexual and the fused perianth forms a small urn that restricts insect access; the gynoecium of the pistillate flower also has a single ovule.

Curiously, in the Melastomataceae, with flowers having stamens with poricide dehiscence, the coleopteran squeezes the anther and éxtracts the pollen.

Psychophilous and palaenophilous flowers have sacarose concentrations in the nectar varying from 1 to 40 percent; the higher percentages are in the palaenophilous. Butterflies have diurnal activity and are usually attracted by small flowers with deep colors and soft, agreeable scents. The corolla is trumpet-shaped with a narrow base and stands in a vertical position; sometimes the flowers are not attractive but the bracts are intensely colored (*Godmania, Mussatia*). The anthesis is diurnal, but the flower usually does not close at night. Moths are the most important lepidopteran pollinators in the Neotropics, especially sphingids; with their elongated proboscis they easily reach pollen. The flowers are white or poorly colored, sometimes red; the tubular or lobulated corolla maintains a horizontal position or is hanging (*Leucocalantha, Tanaecium*). The anthesis is nocturnal, the anthers are versatile, and the scent is sweet and strong.

Many species have the brush syndrome: very small flowers clustered, each with many stamens and stigma above the perianth level. These features attract large pollinators like sphingids. The species pollinated by moths include *Enterolobium cyclocarpum* (Jacq.) Griseb., *Samanea saman* (Jacq.) Merr., *Pseudosamanea guachapele* (Kunth) Harms, *Cojoba arborea* (L.) Britton & Rose., *Abarema adenophora* (Ducke) Barneby & J.W. Grimes, *Albizia niopoides* (Spruce ex Benth.) Burkart, and *Inga* (Fabaceae-Mimosoideae); *Cedrela odorata* and *Guarea* (Meliaceae);

Cordia alliodora (Boraginaceae); and *Quararibea asterolepis* (Bombacaceae). The *Symphonia globulifera* (Clusiaceae) species has been considered ornithophilous and pollinated by hummingbirds, although pollination is carried out by lepidoptera. The hummingbirds remove nectar.

Melittophily is the most common type of entomophily in the Neotropics. Bees may be generalists or specialists and some may detect differences in the pollen availability of diverse flowers. The most important families are the Anthophoridae (*Xylocopa, Centris*) and the Apidae (social bees). *Centris* frequently pollinate flowers such as *Byrsonima crassifolia* that produce abundant lipids but lack nectar. In the Subfamily Meliponidae (Family Apidae) the genera *Trigona* and *Melipona* are important; in the Subfamily Bombinae (Family Apidae) the euglossine bees are the best pollinators (*Bombus*). The morphological diversity found in the group enables pollination of zygomorphic and actinomorphic flowers. Tropical trees pollinated by bees include *Aspidosperma megalocarpon* (Apocynaceae), *Mabea fistulifera* (Euphorbiaceae), *Copaifera camibar* Poveda, Zamora & Sanchez (Fabaceae), *Andira inermis* (W. Wright) Kunth ex DC., *Dipteryx oleifera, D. panamensis* (Fabaceae-Papilionoideae), *Laetia procera* (Poepp.)

Eichler. (Flacourtiaceae), *Couratari scottmori, Eschweilera,* Grias, Gustavia, Lecythis ampla (Lecythidaceae), Cespedesia *macrophylla* (Ochnaceae), and *Qualea paraensis* (Vochysiaceae).

Because Vespidae have high nutritional requirements, especially for carbohydrates, they collect and store nectar. Almost all species of *Ficus* are pollinated by Agaonidae wasps when they oviposit the syconium. Several wasp families may pollinate the *Ficus* species. Because ants (formicophily or mirmecophily) require sugars and proteins, they collect nectar and eat anthers. They are considered nectar thieves and it is impossible to characterise the flower type with which they are associated. Presumably, ants and wasps visit the same kinds of flowers. Flies and gnats visit several types of flowers. The flowers pollinated by Vespidae are small, whitish, greenish, or yellowish, such as *Anacardium excelsum* (Bertero & Balb. ex Kunth) Skeels, *Mangifera indica* L., *Tapirira guianensis, T. mexicana* (Anacardiaceae), *Bursera simaruba* (L.) Sarg.(Burseraceae), *Simarouba amara* Aubl. (Simaroubaceae), and *Goethalsia meiantha* (Tiliaceae).

The group has many pollinators. The cocoa (*Theobroma cacao*) is pollinated by several species of Diptera, and flies pollinate *Monadera myristica* flowers. Birds either pollinate the flowers while eating nectar or are opportunistic. Flowers are usually red, orange, or deep purple, with narrow tubes, thick texture, and no odor. Anthers are exserted or subexserted and the calyx is long and loose. The pollination of *Myrrhinium atropurpureum* (Myrtaceae) flowers illustrates an interesting case.

Flowers are grouped in dichasia and placed in horizontal branches. They lack nectar and scent but the red petals are thick, fleshy, juicy, and sugary; they change in color and shape during the phases of the anthesis (pink, red, purple, black, purple). The stamens have a long filament (15 mm) and each anther contains around 8,000 pollen grains. The style is 12 to 20 mm long. Pollen presentation starts when sugar concentration in the petals is 4 ± 1 percent weight/weight and the petals are red-violaceous in color. This floral syndrome attracts passerine birds, which are the main pollinators; petal color attracts the birds, petal color variation indicates sugar variation, while the long filaments expose the anthers.

In the Neotropics the most specialised pollinators are the hummingbirds (Family Trochilidae), although there are other minor groups such as the Icteridae. Bird pollination is common in epiphytes and the plants of forest clearings and secondary forests. Hummingbirds are very important pollination vectors in the upper levels of the mountain ranges. The birds discriminate among ultraviolet radiations. Most have weak olfactory systems which are, however, sensitive enough to detect odors in the environment that can lead them to food. The anthocyanidic pigments are a typical syndrome of flowers pollinated by birds, although exceptions are found in Australia and New Zealand.

The flowers usually pollinated by hummingbirds may be odorless, and strongly colored, with the red, yellow, and orange colors dominant. The corolla is tubular, sometimes hanging, and lacks a lip or margin turned backward. There are nectaries at the base of the floral tube with an abundance of nectar. The filaments are rigid or fused, frequently stipitate. The position of sexual organs facilitates nectar gathering, pollen collecting, and pollen transference from one flower to another. The anthesis is diurnal. The pollen is usually deposited in the beaks or head feathers of birds. Most pollen grains are light colored (white or pale yellow), spherical, smooth, and puverulent; sometimes the exine has spinules which expedite pollen adherence to bird feathers. In the Myrtaceae, the pollen deposited in bird beaks is smooth, spherical, or triangular and has a viscous brown, blue, or black covering.

Ornithophily is common in the Fabaceae-Mimosoideae (some species of *Inga*), Fabaceae-Papilionoideae (*Erythrina cochleata*, *E. crista-galli*, *E. poeppigiana* (Walp.) O.F. Cook), Rubiaceae (*Psychotria elata*, *Hamelia patens*), and Malvaceae (*Malvaviscus arboreus*).

Hummingbirds pollinate these species. *Vochysia* flowers, usually pollinated by bees and butterflies, are sometimes pollinated by small hummingbirds. Birds are not always specific pollinators; for example, several passerine birds are diurnal pollinators of *Mabea fistulifera* (Euphorbiaceae).

Nectarivorous bats are the most important group of mammalian pollination vectors. They are nocturnal, blind to color, and have a well-developed olfactory system. Because their sonar is weakly developed, bats have trouble flying in foliage. They may use their claws to hang from flowers or fly in front of the flowers like hummingbirds. The tongue is long and well adapted to extracting nectar from the flowers. The bats feed on flowers that produce abundant pollen and sticky or mucilaginous nectar.

The corolla is usually wide distally and the nectar very accessible. The flowers pollinated by bats usually stand out in the tree crown, have a special position (cauliflory), are hanging (penduliflory), or have long pedicels. The corolla is campanulate or infundibular, whitish or creamy, greenish, purpuraceous or pinkish, and the flower has glandular tissues exuding unpleasant odors (butyric acid); the anthesis is nocturnal. They can be individual or be grouped in inflorescences. The pollen is transported from one flower to another in the bat hairs, which have scales resembling those of the bee abdomen.

Plant families such as Bombacaceae are pollinated primarily by bats (e.g., *Bombacopsis quinata* (Jacq.) Dugand, *Ceiba* pentandra (L.), Pachira aquatica Aubl., Ochroma pyramidale (Cav. ex Lam.) Urb.). A high number of Bignoni-aceae (Crescentia cujete L., Amphitecna sessilifolia), Capparidaceae (Capparis), Caryocaraceae (Anthodiscus chocoensis, Caryocar *costaricense* Donn. Sm.), Euphorbiaceae (*Mabea*), Fabaceae-Mimosoideae (*Inga leiocalycina, Parkia pendula* (Willd.) Benth. ex Walp., and Fabaceae-Caesalpinioideae (*Peltogyne purpurea* Pittier), are also pollinated by bats. Both bats and sphingid insects pollinate some species, such as *Capparis pittieri.* Wind pollination is difficult to detect because many flowers are also visited by diverse insects or other pollination vectors. Pollination by wind and insects (ambophily) is very common Silberbauer.

In general, the wind is the pollination vector of species with unisexual flowers that have reduced or no perianth and lack substances that attract birds and other animals. The anthers are exposed and have long filaments. Air-dispersed pollen grains are small (± 16 to 50 µm), light, and spherical; the exine is thin, sometimes reticulate, striate-reticulate, echinate, scabrous, psilate, sometimes rugose, geminate or discontinuous, puberulent and nonadhesive. The pollenkitt covering is restricted to the exine arcs. Pollen produc-tion is abundant in the species. Conditions that favour wind pollination include the absence of rain, low relative humidity, and good air movement (wind pollination is frequent in dry tropical forests).

In the Neotropics, several families have species that are partially or exclusively wind-pollinated. These species include *Astronium graveolens* Jacq., Tapirira guianensis, T. mexicana (Anacardiaceae), Alnus acuminata (Betulaceae), *Terminalia amazonia* (Combretaceae), Bernardia, Croton (Euphorbiaceae), Quercus (Fagaceae), Xylosma *intermedia* (Flacourtiaceae), *Alfaroa costaricensis* (Juglandaceae), Maclura tinctoria, Sorocea, Trophis racemosa (Moraceae), *Forestiera* sp. (Oleaceae), *Agonandra racemosa* (Opiliaceae), *Zanthoxylum* (Rutaceae), *Ulmus mexicana* (Liebm.) Planch. (Ulmaceae), *Myriocarpa longipes*, and *Urera caracasana*

(Urticaceae). Anemophily is also present among gymnosperms such as *Podocarpus guatemalensis* (Podocarpaceae) and *Pinus* (Pinaceae).

The pollen dispersed by water has different shapes: spherical grains surrounded by mucilage; ellipsoid grains, in linear tetrads enclosed in mucilage tubes; or filiform grains lacking exine, with an intine layer resembling that of the terrestrial pollen. The tropical trees of lowlands and coastal plains have low population densities and are pollinated primarily by animals. Pollination mechanisms, pollinators, and sexual systems vary among angiosperms. Insects are the most common pollinators and large and medium-sized bees are the most frequent vectors, followed by moths, wasps, other insects, and small bees. The pollination vectors of almost one-half of the plant species forage over a wide range. The highest diversity in pollination systems occurs in the subcanopy, where hummingbirds and sphingid insects are the most active. Bats and wind are minor pollinators. Pollination systems are more uniform in the canopy, where bees and small insects dominate.

Most tropical trees have incompatible bisexual flowers (perfect or hermaphrodite). Some are dioecious (pistillate and staminate trees) and cross-pollination is obligatory; however, the strength of the incompatibility barrier varies in a population. In forest management, the degree of individual crossing is vital to determining the genetic quality of the seed. The degree of endogamy depends not only on the genetic tendency to individual self-pollination but also on the spatial distribution of the individuals.

Dispersal from the site of origin promotes gene displacement throughout a population or to new populations. In species with cross-pollination, the paternal genes move twice in each generation, first during pollination and second during seed dispersal. The maternal genes move only once; therefore, the paternal genes move farther in each generation. The pattern of dispersal contributes to structure population, to potential genetic drift, and to natural selection. The occasional gene transference from one population to another conspecific population is important to maintaining the genetic diversity of the recipient population. The invasion and colonisation of new lands permits the formation of monospecific populations. Natural or artificial monospecific populations are at high risk for diseases and infestations of parasites.

2

Seed Collection

There is a great variety of methods and equipment available for collection of fruits and the choice depends on a number of factors which are summarised as follows:

1. Relative size and numbers of the natural dispersal units and of the units which can be conveniently collected by man. In the case of 1 – 3 large seeds inside a dehiscent or indehiscent fruit (e.g. *Aesculus*, *Tectona*), collection can be done most easily by awaiting natural fall of seed or fruit and collecting from the ground. At the other extreme collection from the tree of fruiting heads of *Adina cordifolia* at 200 per kg is the only practicable way to collect the seeds; at 11 million per kg, it would be impossible to collect them after dispersal.
2. Characteristics of the fruit: size, number, position and distribution of fruits; resistance of peduncles to shaking, pulling, breaking or cutting; interval between ripening and opening.
3. Characteristics of the tree: diameter, shape and length of bole, bark thickness; shape of crown; size, angle, density and resistance to breakage of branches; density of foliage and depth of crown.

4. Characteristics of the stand: distribution and stocking of trees (e.g. isolated trees, open or dense stand); density of understorey and ground vegetation).
5. Characteristics of the site: slope, accessibility.

The various collection methods may be classified into the following: (a) Collection of fallen fruits or seeds from the forest floor, (b) Collection from the crowns of felled trees, (c) Collection from standing trees with access from the ground, (d) Collection from standing trees with access by climbing and (e) Collection from standing trees with other means of access.

Collection of Fallen Fruits

Natural seedfall

Collection from the forest floor of fruits which have fallen after natural ripening and abscission is common practice with a number of large-fruited genera. It is cheap and does not require as highly skilled labour as, for example, climbing; school children or casual labour may be used. Fruit size is very important as the larger the fruit, the easier it is to see and pick up by hand. Temperate genera commonly collected from the ground are *Quercus, Fagus, Castanea* and tropical genera include *Tectona, Gmelina, Triplochiton* and several genera among the dipterocarps.

The main disadvantages of collection from natural fruit shedding are the risks of collecting immature, empty or unsound seeds, of seed deterioration or premature germination if collection is delayed, and uncertainty in identifying the mother trees from which seed is collected. Seeds in the first fruits to fall naturally in the season are often of poor quality. In Thailand shedding of teak fruits starts in March, but observations have shown that the most viable fruits are shed in the latter part of the season, so collection is usually postponed until April. Clearing the forest floor of vegetation and debris, including old or prematurely fallen fruits, and/or spreading out sheeting of light canvas, calico or plastic, to catch the seed, can greatly facilitate collecting efficiency. If carefully timed, this operation will also eliminate much of the risk of collecting empty or non-viable seed.

Sound fruits should be gathered as soon as possible after they have fallen, to avoid damage or losses from insects, rodents or fungi and premature germination. This is of particular importance in the moist tropical forest. Observations have indicated that many of the seeds of the more important dipterocarps lose their viability within a few days of shedding and studies on *Shorea platyclados* in Malaysia demonstrated that seed lots collected from the ground included considerably more defective seeds than lots collected from the standing tree. Collection from the ground must, therefore, be perfectly timed with seedfall.

In the Jari region of Amazonia in Brazil, Woessner and McNabb found that collection of green or yellow fruits of *Gmelina arborea* from the forest floor gave the best results in operations for collecting about 10,000 kg of seed a year. They could be stored temporarily in sacks during transit from the field to the fruit processing depot, without serious loss of viability. Older brown or black fruits ferment and heat in the sacks and rapidly lose viability. Collecting teams are instructed to collect only the fresh green and yellow fruits. 50 kg of fruits can be collected per man in an 8 hour working day and yield about 3 kg of dried stones. Similar results have been obtained in Malaysia where green and yellow fruits collected from the ground gave over 90 % germination, but brown fruits, on the other hand, produced only 53 %.

Seeds of some hard-coated species may remain viable on the forest floor for years, especially in temperate conditions. In Hungary seed of *Robinia pseudoacacia* is collected from the forest floor under 30-year-old stands in the Pusztavacs forest district. A special machine screens the top 10 cm of soil and yields about 770 kg of seed per hectare, which is the equivalent of the yield of approximately 10 seed years. Even in the tropics viable hard-coated seeds may be obtained by screening the soil below the mother trees. This has been done in Malaysia for *Parkia javanica* and *Intsia palembanica* of which the seeds are large enough to be picked up by hand. Where the seeds are smaller e.g. in *Albizzia falcataria*, sieving with a wire mesh may be more practical.

The identity of the mother tree is often uncertain when fruits are collected from the ground. Isolated trees present no problem in this respect (though they may be undesirable parents because of the risk of selfing), but much mixing of fruits can occur in dense monocultures with interlocking crowns. This is of no concern in collecting commercial quantities of seed, provided that the genetic quality of the stand is average or above average.

For research and breeding purposes, it is often necessary to maintain the identity of the mother tree of each seed lot. In such cases it is advisable to clear the grond of already fallen fruits and to accelerate the fall of new fruits by shaking, beating or cutting off branches, or climbing and picking fruits in the crown. A compromise solution, suitable for commercial collections in unimproved stands containing a mixture of good and bad phenotypic trees, is to collect fruits only below the better seed bearers and within half the radius of projection of their crowns.

Manual Shaking

If fruits are easily detached but natural fruit fall is insufficiently concentrated in time, fruit fall may be induced by artificial means. Trunks of small trees and low branches may be shaken directly by hand. Higher branches may be shaken by means of a long pole and hook or by a rope. This method has produced good results in *Cordia alliodora* and *Cedrela* spp., as it facilitates rapid collection of seed with good viability as soon as visual inspection shows that the fruits are mature.

Use of a rope involves an initial operation to pass the rope over the branch to be shaken, which is described by Robbins et al. The same method is used to hoist a saw or pulley into the crown. A thin line is attached to a weight, which is projected over the branch by hand or by catapult.

For higher branches the line may be attached to an arrow, which is shot from a bow, or to an iron rod shot from a calibre .22 rifle. A light nylon line such as a fishing line of 50 lbs. (23 kg) breaking strain is suitable and the weight or projectile used should be heavy enough to drop to the ground pulling the line with it over the branch.

Care must be taken to ensure that the line unreels readily without getting entangled, for example by using an appropriate fishing reel. Once the end of the line has reached the ground, the weight or arrow can be detached and a nylon cord of 3 – 4 mm attached instead; the line is drawn back over the branch, pulling the cord with it. The two ends of the loop can then be pulled together to shake the branch. The cord should be positioned towards the end of the branch where it will have the maximum shaking effect and not close to the bole where the branch is thickest.

In New Zealand a 30 kg-pull fibreglass long-bow firing arrows 0.8–0.9 m long and 40 g in weight has been used to project a line of 5.5 kg breaking strain over a branch. This is used, successively, to pull up a 20 kg line and a 180 kg line with pulley. In Canada both a .45 calibre gun firing a steel rod of 270 g weight and a .22 calibre gun firing a cylinder of 230–300 g weight and 2 ½ inch diameter have been used successfully up to heights varying from 20–50 m. Blank cartridges were used. The line attached to the projectile was 32 kg monofilament fishing line, which was used to haul up a 320 kg plastic twine carrying a branch breaking or cutting device.

Mechanical Shaking

Mechanical tree shakers were developed originally for fruit and nut orchards but since about 1965 the technique has been used for certain forest trees, particularly for the southern pines in the USA. The machines are expensive, need flat ground for efficient use, and experienced operators are essential to avoid excessive damage to the trees. Many cones are removed by a few seconds of shaking, but longer shaking breaks off pieces of tops and limbs. Tree shakers have no role in diffuse collecting operations in natural forest, but will probably continue to be used in intensively managed seed orchards or seed stands of a limited range of species.

The American Shock Wave tree shaker is mounted on a short wheel base truck chassis, equipped with an automatic transmission. It has a padded clamping device mounted on the extreme end of a 6 metre boom capable of clamping the trunk of a tree up to 90 cm diameter. A shake pattern is developed by counterrotation unbalanced weights in the shaker of varying frequency from 400 – 4000 cycles per minute.

The tree is generally clamped about 3 m above the ground and one 15-second shake is sufficient to remove about 80 percent of the cones from *P. elliottii*; but *P. taeda* and *P. echinata* are more difficult and a good operator may often remove only 25 – 30 percent of the cones after prolonged shaking. Ripe *P. elliottii* cones require a force of about 2 kg for removal whereas *P. taeda* cones are detached only by a force of 20 kg or more. An unsuccessful attempt to reduce the force needed to remove the cones of *P. taeda* by the use of abscission inducing chemicals is reported by McLemore. The repeated shaking of the more difficult trees can cause bark rupture and breakage of the leading shoot.

With five mechanical shakers the Louisiana Forestry Commission harvested cones from 34,680 *P. elliottii* trees in twenty days. Three-quarters of the trees released 85 percent of their cones during shaking periods from 6 – 30 seconds. Performance in one hour exceeded that of a climber in a week. Results reported by McLemore and Chappell showed that the mechanical shaking of *Pinus elliottii* trees by trained operators did not harm future cone production, tree growth or vigour over the following four years.

Mechanical tree shakers are now widely used in the southeastern USA for harvesting seed in seed orchards of pines. In the case of species in which ripe cones can be easily detached, such as *P. elliottii* and *P. palustris*, trees are shaken in the period after cones have reached maturity but before they open, and the cones with their contained seeds are collected from the ground. In the case of species with persistent cones, such as *P. taeda* and *P. echinata*, shaking is deferred until the cones have opened and the objective is to shake out the seeds from the cones.

In the U.S.S.R. output from the VUS-2 vibrator working on *Pinus sibirica* was reported to be 10 – 20 times that obtained by manual methods. More recent experience with a tree-shaking machine developed in Soviet Central Asia has shown that 90 – 100 % of fruits of *Juglans, Malus, Prunus, Fraxinus* and *Gleditsia* can be collected with little stem damage. The machine is a grab-type vibrator mounted on a hydraulically operated arm on the 3-point linkage of a tractor. Optimum shaking time is 10 – 25 seconds and optimum vibration rate for most species 1000 cycles per minute. Italian CECMA vibrator machines, developed primarily for harvesting the olive crop, have also been used successfully for collecting cones of Mediterranean pines.

Dispersed Seed Collection

Although collection from the ground is most often used for fruits, it can also be used for seeds dispersed after the cones or fruits have opened. The seeds of southern pines in the USA, e.g. *Pinus elliottii* and *P. taeda* have a very short period between reaching maturity and being dispersed, and a number of methods have been evaluated for collecting loose seeds as they are dispersed. In addition to the use of sheeting spread on the ground, mentioned previously, they they include polypropylene netting round the crowns, funnel-shaped wooden frames covered with cloth or polyethylene and attached to a central hub surrounding the stem, and sheeting or nets raised on poles above the ground. Earlier experience was not very successful, as rarely more than 50 % of the available seed crop was recovered. If seed is borne mainly on or near the outside of the crown, much of it falls outside the spread of a single tree catching unit. If sheeting or netting must remain in place for an extended period of natural seed fall, it is liable to get damaged from weather and a proportion of the seeds are lost to birds and animals.

More recently a net retrieval system, for use in seed orchards in southern USA, has been developed jointly by the Georgia Forestry Commission and the Missoula Equipment Development Center. It shows considerable promise and is increasingly preferred to vacuum harvesters in *P. taeda* orchards. Netting used is a polypropylene plastic fabric manufactured for carpet backing. It is light, durable and available in a range of dimensions. For *Pinus taeda* a width of 16.5 ft (5m) and a weave count of 6 × 8 per square inch (approximately 2 × 3 per cm^2) is used.

If carefully handled, the expected life is 10+ years. Netting is laid in the orchard several weeks before seed fall and trees are shaken by mechanical shaker to dislodge seeds from the cones. The power take off from a wheeled tractor provides the power (less the 30 h.p.) needed for (a) Mechanical winding of the net onto a roller and (b) Mechanical separation of seed from the twigs, leaves and cones which also fall to the ground as a result of the shaking operation. Care must be taken in separating the netting from the grass blades which tend to go through it, before starting to wind in the netting.

Gathering of fruits on the ground is usually done manually but may be assisted by using a simple hand tool, such as a long-handled rake with interchangeable heads having different numbers of, and spacing between, tines. Attempts have been made to develop mechanical or vacuum sweeper methods to pick up seeds or fruits. Tests have also been made of a mechanical harvester, with the sweeping action of a rotating drum fitted with a myriad of attached rubber fingers which pick up seeds with separation from foreign debris. These types of machine are best used in conjunction with a tree shaker, which ensures that a substantial quantity of seeds are on the ground at each sweeping operation.

Mineau described the successful use in France of a vacuum sweeper type of machine, operated by compression from a tractor engine, for collection of *Fagus* fruits. The machine is compact and weighs 450 kg. A Dutch vacuum seed harvester has been used successfully for the collection of *Quercus* acorns and is reported to be a cheaper method in the Netherlands than either hand-picking or the use of a tree shaker and tarpaulins.

In the USA, Hallman summarised the advantages of the vacuum sweeper as: it

1. Extends the harvesting season from about 2 weeks to 2 months.
2. Eliminates the need for ladders and lift trucks, permitting orchard managers to let their trees grow taller because no climbing is involved, therefore extending the useful life of the trees.
3. Reduces collection costs compared to hand collection.

The disadvantages of the harvester are that:

1. The orchard floor requires extensive preparation for the machine to operate well.
2. The harvester is noisy and produces large quantities of dust.
3. It does not operate well when the ground is wet.
4. The harvester has had a number of mechanical problems that have been largely corrected, but minor design changes still need to be made.

The disadvantages are now considered to outweigh the advantages, and the net retrieval system, described on p. 53, is preferred for collecting *P. taeda* seeds in orchards in the southern USA.

In view of the capital cost of this type of equipment and the emphasis placed on the immaculate preparation of the seed orchard floor which is needed for its operation, its use in developing countries is likely, especially if social considerations favour the use of labour-intensive manual methods as a means of increasing employment.

Animal Caches

Animals sometimes gather together cones or fruits as a food supply and these caches may be raided for the seed, but this source of seeds is confined to limited areas. Squirrel caches are an important source of coniferous seed in western North America. Squirrels usually locate their caches year after year in the same places.

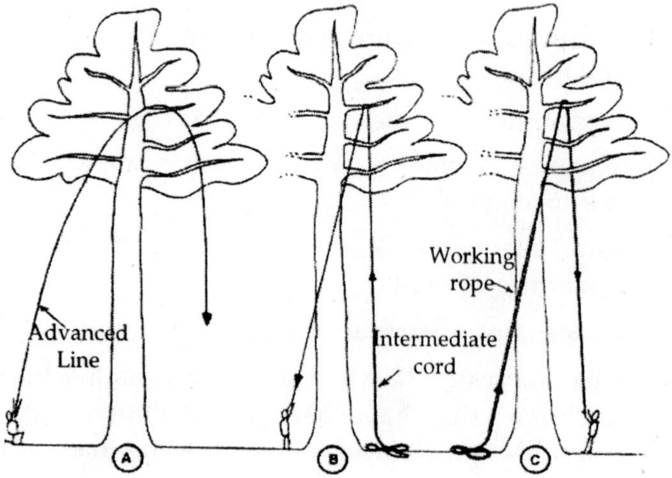

Contd....

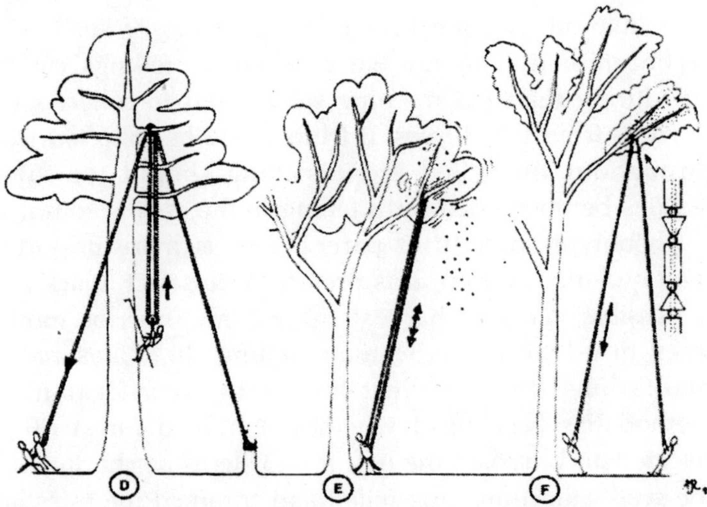

Figure 1. The Advanced Line Technique (A)-(C) show stages from shooting the advanced line to drawing up the working rope (D)-(F) show possible uses of the working rope, (D) with block and tackle to pull man into crown (E) for branch shaking (F) to sever branches by means of flexible saw.

Typically, they are found in damp areas near springs, small creeks or marshes, on northern exposures, and in decayed wood or duff or around old fallen trees. A single cache may contain from a few cones to many bushels. Fresh cones on the ground are a sign of squirrel activity; piles of cone scales and cores may indicate a nearby cache. Caution should be exercised in collecting seeds or cones from caches, because of the danger of infestation by pathogenic fungi which may reduce germination.

Ants sometimes gather seeds together and in North Africa they have been observed to accumulate large piles of *Acacia* seeds. Any seeds collected from rodent or insect caches should be tested for soundness by cutting test or other means.

Seed Collection from Felled Trees

One method of collecting large amounts of seed is to synchronise it with normal commercial fellings carried out during the seed ripening season and to collect seeds or fruits from felled trees. If fruits are to be collected from throughout the felled crop, picking should for safety reasons be postponed until felling in the area is complete. If phenotypic quality of parent trees is more important than quantity of seed, it is preferable to select, mark and, if possible, fell and harvest fruits from superior mother trees in advance of the main felling. In *Pinus radiata* plantations due for clear-felling in New Zealand, a method has been used whereby a trained forest officer selected and marked the best 8 – 13 stems per hectare and the seed collection gang felled and trimmed the trees and collected the cones, so as to leave no obstructions for the subsequent felling of the remaining trees.

Collection of fruits from early thinnings should be avoided, since it is difficult to judge phenotypic quality correctly at that age. Felling crowns into existing gaps is advisable in order to facilitate recovery of cones. It is essential to confine collection to the season when seeds are mature; adjustment of felling dates to coincide with seed ripeness should be possible wherever the same authority is responsible for both the felling and the collecting activities, e.g. national forest services operating within state forests. Hand picking of cones or fruits in the fallen crowns is common practice, assisted by rakes, hooks or machetes. Small cluster-type cones, such as those of *Thuja* and *Tsuga*, may be harvested by cutting off the cone-laden branch tips and pulling them through a cone stripper. The stripper contains a series of teeth, similar to a rake, which are set sufficiently close together to remove the cones.

In practice collection in clear-felled areas has proved little if any cheaper than collecting from standing trees by a well-trained team of climbers, at least in north-temperate conifers. The tangle of fallen stems and crowns and the dispersal of some cones during felling greatly reduce productivity. Where operations are speedy and closely controlled, collection after the boles have been trimmed and removed but before the lop and top has been piled and burnt may prove the most efficient method.

Felling of selected individual trees specifically for seed collection may be necessary in areas where commercial fellings are not practised, for example when relatively small quantities of seed from a few trees are needed for provenance testing or other research purposes. Such special fellings should be avoided whenever possible, both because the bole is wasted and because the tree is lost as a future seed source, but they are sometimes inescapable in the case of tropical high forest species which are very difficult to climb and if a seed collecting expedition is severely limited in time. Felling tall but unbuttressed trees is usually a lot quicker than climbing them. The collection of fruits from wind-thrown trees is generally undesirable, as little selection can be applied and there may be a bias towards trees with characteristics which pre-dispose them to wind damage.

Seed Collection from Standing Trees

By Hand

In the case of shrubs or low-branched trees, fruits can be picked directly from the branches by the collector while standing on the ground. Examples are *Crataegus, Sorbus* and *Ilex* spp. in temperate zones, the smaller acacias and mallee eucalypts in Australia and many of the small drought-resistant species of the arid and semi-arid zones. Smaller fruits are generally harvested directly into a basket, bag, bucket or other container held or worn by the picker.

Cutting, Breaking and Sawing

For branches out of arm's reach a variety of long-handled tools is available to enable the collector to reach the fruits from the ground. A pole and hook may be used to pull branches down within reach. Long-handled rakes, saws, chisels, hooks or pruning shears are used to pull off or sever individual fruits or fruit-bearing branchlets. Light rigid bamboo, aluminium or plastic poles 4 – 6 m in length are common. In order to reach beyond the 6 – 8 m range of single poles, multistage telescopic poles with a shear on the end have been developed. Robbins et al. have noted that, in some species, fruits or cones on the lowest branches may yield little seed, because of lack of pollination in that position, and that it is therefore preferable to collect fruits from at least half way up the crown. Ability to use long-handled tools efficiently from the ground is much affected by the density and form of the crown in individual trees.

A rope can be thrown or pulled over a branch as already described, but used to break off the seed-bearing branch rather than to shake it. A thicker rope is needed than for shaking. The method is not recommended for general use. It damages the tree, allows access to pests and diseases and, in the case of pines and other species which take two years to mature their seeds, destroys the next year's seed crop while collecting the current year's.

Several types of *flexible saw* have been used successfully to sever branches from the ground. One model, described by Anon. 1979, consists of a 3 ft. long flexible cutting cable fitted with precision-set carbon steel teeth and two 35 ft. polypropylene control lines. A sand-filled safety weight is used to project one of the control lines over the branch. An earlier model, the "commando saw", now no longer in production, was effective in servering eucalypt branches in Australia. Two operators could bring down branches up to 20 cm in diameter quickly and easily.

The method is not applicable to trees with acutely angled branches such as *E. tereticornis*. The other limitation depends on the efficiency of projecting the line over the desired branch. Rigid saws can also be used to sever branches. Sweney and Jones describe a method used in New Zealand in which a pruning saw or bow saw is attached by lugs to a 180 kg line running through a pulley previously raised into the crown. The pruning saw is used for small branches <2 cm diameter, while the bow saw will cut through a 10 cm branch in 5 minutes' sawing.

Use of Rifle

Another method of severing seed bearing branches is to shoot them down with a large calibre rifle. The method was successfully used to shoot out the tops of *Picea glauca* trees in seed production areas in north eastern USA. Not only was the topping of the trees found to be less expensive than climbing but the cones could also be collected at the best stage of development because of the short time in which the operation could be completed. More recently shooting off branches or tops from a helicopter has yielded promising results in Canada.

In Australia the collection of small samples of eucalypt and *Araucaria* seed from tall trees has been accomplished efficiently using a .222 or .243 or .308 calibre rifle with x4 telescopic sights. Branches up to 15 cm diameter could be brought down. "Pointed soft point" ammunition is more effective than "hollow point" when used with a .308 rifle.

A disadvantage of the rifle method is that very strict safety precautions must be observed. There are limitations to where a rifle may be used, for example not near roads or built-up areas. Also crowns of some species such as *Araucaria* and *Picea* may be considerably damaged by this technique.

In shooting down branches it is usually necessary to steady the rifle on a tripod or to rest the stock against a tree or the side of a vehicle. A clear line of sight is required and this can be a limiting factor in dense forests. It is usually best to shoot at right angles to the branch and to sever the bark on the underside with the first shot to avoid branch hang-ups. The bark on the upper side is then cut and finally shots are placed at intervals across the branch. It is important to select branches which will fall unobstructed to the ground. Horizontal branches are more readily detached than ascending branches. The shots should be positioned to take advantage of branch leverage. The method is best suited for collecting research quantities of seed from a heavy seed crop clustered on branches or tops too inaccessible to be conveniently reached by other means.

Seed Collection from Standing Trees with Access by Climbing

There is a limit to the height to which long-handled tools can be used for collecting seeds or fruits from the ground. Near that limit the operation consumes much time and energy but produces little seed. For tall trees which cannot be felled, therefore, climbing is often the only practical method of collecting. Some men are excellent natural climbers, while good training and good equipment can render collection by climbing an efficient and safe, albeit energetic, operation. For convenience the operation may be described under the following subheads: (a) Climbing into the crown by way of the bole, (b) Climbing into the crown directly, (c) Climbing and picking of fruits within the crown.

Climbing into the Crown by Way of the Bole

Climbing with minimum equipment: Climbing without mechanical aids is practised in a number of countries. In the Philippines some seed collectors climb barefooted or with the help of a rope which ties both feet together and presses them against the trunk of the tree. Other modifications are for the climber to cut successive notches in the bole with a hand-axe to support his feet, or to hammer in a series of iron spikes about 20 cm long which are later withdrawn for re-use as he descends. Both of these methods are physically exhausting, whether or not a safety belt is used, and do some damage to the tree. Climbing tall branchless boles with hands and feet involves a considerable safety hazard and the risks may tempt climbers to prefer collecting from the most easily climbable trees which are often silviculturally the least desirable. It is preferable to introduce one or other of the special climbing aids now available.

Climbing irons or spurs, which are attached to the climber's boots, offer a light and inexpensive means of safer and more efficient climbing, if combined with safety belt, strap and line, safety helmet of glass fiber and heavy leather gloves. The lightness of the spurs (less than 1 kg a set) makes them particularly suitable for use in inaccessible stands in roadless country, where all equipment must be carried on foot. They have been found to be the most efficient method for climbing trees of *Pinus kesiya* and *P. merkusii* in Thailand and for *P. caribaea* and *P. oocarpa* in Honduras and are in common use in many countries, especially for conifers.

There are a number of different types of climbing irons but basically they consist of a forged iron arm and connecting piece which terminates in a pointed spur. The iron must be fastened securely by a leather strap to the foot gear and sometimes to the leg of the climber. The spur may be of varying length but it is an advantage if the point does not extend beyond the sole of the boot, so that the climber can walk on the ground without difficulty. The optimum length of spur depends on the type of bark. 5 cm spurs are suitable for barkless telephone poles and thin-barked trees and are recommended for most species in Canada, while 9 cm spurs are better suited to species with soft, thick bark. Climbing irons should not be used when the bark is frozen and are not particularly safe on scaly bark.

Detailed guidance to the fitting, use and maintenance of treeclimbing spurs in Canada is contained in Yeatman and Nieman. The following description of climbing with spurs closely follows their account. The climber ascends the tree using a safety belt, with a safety strap or safety chain passed around the stem and hooked to the belt. A safety line is tied to the belt and two or more carabiners are clipped onto one of the belt rings. When climbing the stem, the climber must ensure that the spikes are well into the wood of the tree by keeping the knees out from the stem when setting the spur. The lower leg and ankle must be kept at a fair angle to the stem to prevent slipping and gouging the bark. The weight is kept on the feet spaced 15 – 20 cm apart and the centre of gravity away from the stem. The hands and arms are used to balance by holding the safety strap firmly in both hands, rythmically pulling the body towards the tree, moving the strap as the weight on it is lifted and tightening it in the new position as the body moves back. The pull on the safety strap is from the arms in ascending and it should not be transferred to the safety belt except when the climber is in a resting position. When the safety strap is tight, each foot is moved in turn and the weight transferred to the other foot. The safety strap is never unclipped except to bypass branches too heavy to break off. A second safety strap or carabiner with safety rope should be fastened above the obstructive branch before the first strap is unclipped. When sound branches are reached, preferably at the bottom of the live crown, the safety line is passed through a carabiner fastened by rope above the first branch, the safety strap is unclipped, and the climber works his way into the live branches.

The main disadvantage of spurs is the damage they do to the bark, particularly of thin-barked species. If climbing is only occasional, this should not be excessive, but frequent climbing of the same tree, e.g. for pollination and seed collection in seed orchards, is liable to cause an unacceptable degree of damage; other climbing methods should then be preferred.

Ladders: For heights from about 8 to 40 metres, vertical scaling ladders in several sections provide a safe and convenient means of climbing the bole to the live crown. They can be made of a variety of materials including wood, aluminium, magnesium alloy etc., but each section must be light enough to be easily pulled up by the climber. The legs of the bottom section can be placed on adjustable platforms for greater stability. The length of each section varies between 1.8 and 3 m and its weight should not exceed 3 – 4 kg.

The bottom one or two sections of the ladder are set up parallel to the tree stem with the bracket at the top against the trunk. The climber ascends with his safety strap around both the trunk and the ladder until his shoulders are level with the top of the ladder and then fastens it to the trunk by a rope or chain. Subsequent sections are pulled up by rope and fitted into the section below. Each section is climbed and fastened to the tree in turn. The same procedure is used for climbing from bole into crown as described above for climbing irons. With the lighter sectional ladders the climber can carry two sections, each 2 m long, attached to his safety belt. If the bottom two sections of 3 m each are fitted together and raised into position from the ground, this means he can erect a total of 10 m before needing to use the tool line to haul up additional sections.

Sectional ladders may be designed as "one-legged" or "two-legged". Two-legged are more common. The "one-legged" ladder comprises a central support with small bars fixed as rungs alternately on either side and is attached by chain or rope to the tree. The central support is usually made of steel or wood, so these ladders are not very different in weight from "two-legged" types. They have the advantage of being easier to lodge on uneven ground and they are also more easily manoeuvred between the branches and on very sinuous stems.

Sectional ladders can be used without any risk of damage to the tree. They can be awkward to handle in stands with dense canopy or undergrowth and are much heavier to carry than climbing irons, especially if long clear boles impose the need for many ladder sections. They are also more expensive. They are therefore of limited use in inaccessible roadless country, but are ideal in seed orchards or plantations in flat topography.

The Swiss tree bicycle or "Baumvelo" is a device for climbing tall straight trees which are branch-free to the live crown. It is lighter to transport than sectional ladders but heavier than climbing irons. It does no damage to the tree. It is suitable for use on stems with diameters ranging from 30 – 80 cm. Olesen found it very useful for climbing pine trees in Mexico. In the UK it was found particularly useful for conifer species with large cones e.g. *Pinus, Picea, Pseudotsuga,* in which the collector seeds to move from branch to branch to pull off the cones, rather than remaining for long in the same part. Its use in that country is now, however, restricted mainly to research collections, since the increased area of mature plantations allows bulk seed collections to be made easily from felled trees and ladders are more convenient for seed orchard collections.

The tree bicycle contains two separate units, one for each foot. Each unit consists of an arm (longer in the upper, shorter in the lower unit) to which is fixed a rubber supporting block which rests against the trunk. The lower end of the arm carries a stirrup or pedal with strap and quick release clips, which holds the climber's foot. The upper end is attached to a steel band forming a circle of adjustable diameter around the stem. The tree bicycle is used in conjunction with safety harness or belt, support chain or strap, safety line, safety clips or carabiners and nylon ropes.

Operation of the tree bicycle is well described by Seal et al. and Yeatman and Nieman. The steel band should not fit too tightly round the stem, which would hinder them sliding upwards. They will grip properly as long as the toe of the climber's boot does not touch the trunk when his full weight is on the stirrup; when that happens it is necessary to unclamp, shorten the loop and reclamp the band on each foot in turn.

After securing the bands round the base of the tree, the climber fits his feet into the stirrup, closes the quick-release clips and fastens the straps. He also passes his safety strap or chain round the stem, adjusting its length in accordance with stem taper. The climber ascends, putting his weight alternately on one or the other stirrup and lifting in turn the free foot to raise the loosened band. He pauses to adjust the bands, one at a time, to allow for the taper of the tree. Obstructing branches must be pruned flush with the stem as they are encountered.

The climber continues to ascend until the upper steel band meets the lower living branches of the crown. He then reaches into the crown to fit a nylon safety rope to hold the safety line and proceeds to "park" the tree bicycle.

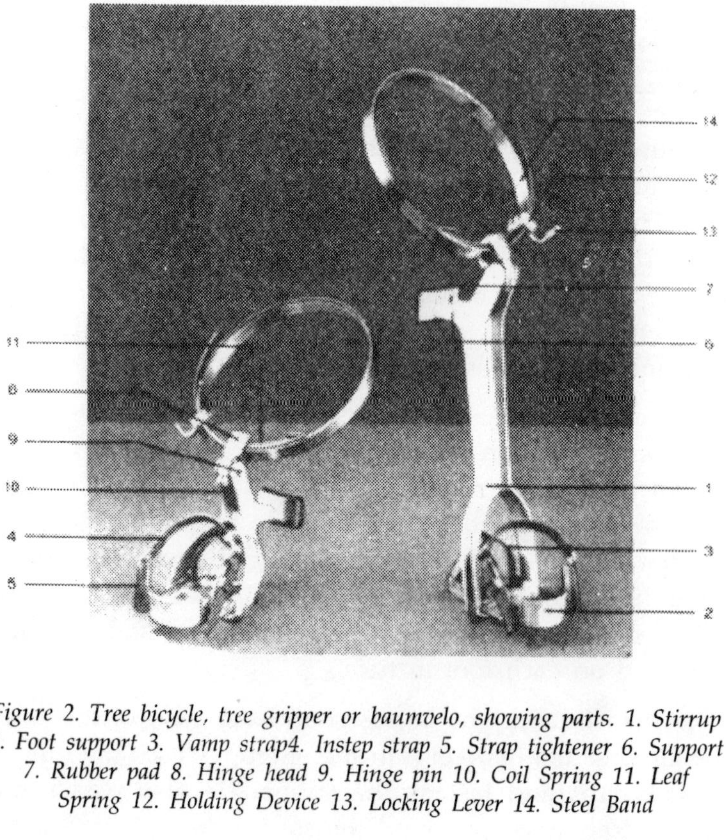

Figure 2. Tree bicycle, tree gripper or baumvelo, showing parts. 1. Stirrup 2. Foot support 3. Vamp strap4. Instep strap 5. Strap tightener 6. Support 7. Rubber pad 8. Hinge head 9. Hinge pin 10. Coil Spring 11. Leaf Spring 12. Holding Device 13. Locking Lever 14. Steel Band

The essential operation is to tighten the lower band of the bicycle so that it grips the stem tightly even when there is no weight on it; there is then no risk of the bicycle slipping down the stem out of the climber's reach. The climber opens the ankle clips on both stirrups, frees his feet from the straps, unhooks his safety strap and climbs into the crown.

The tree bicycle provides an extremely safe means of climbing straight branchless trees without damaging them and is lighter and more portable than sectional ladders. Some practice is needed but most men become competent and quick in its use within a few days. Its main disadvantages are the cost, the fact that its use is limited to a certain range of diameters and that, unlike ladders and climbing irons, it requires the bole to be pruned of branches all round the circumference up to the living crown. Where regular climbing of the same trees is foreseen, however, as in seed stands or seed orchards, the cost of thorough initial pruning is fully justified. It is advisable to work always with two tree-bicycles within call of each other, because a man in difficulties on a tree at a height greater than can be reached by a ladder can *only* be reached on another tree-bicycle, or by climbing irons.

Direct Climbing into the Crown

Ladders: Access to stout lower branches in the crown may be obtained directly from the ground or by ladder, provided that the branches are not too high. Free-standing domestic step-ladders or taller tripod ladders have the advantage that they do not need to be rested against the tree; they are awkward to handle in dense stands but are suitable for collection in seed orchards or heavily thinned plantations where trees are widely spaced. Tripod ladders up to about 6 m are available but must be guyed with ropes on each side for stability.

General purpose ladders are available in wood or aluminium alloy and may either be constructed as a single unit or may be extending, when they consist of two or more sections constructed so that the height may be varied by a relative sliding movement of the sections. Single-section ladders designed for fruit-picking have a splayed base fitted with moulded rubber feet or metal spikes to give a firm grip on soft ground and can be used to heights of 8 – 11 m.

The Finnish Tarra-tikkaat ladder, designed specially for forestry, is similar but incorporates an aluminium ring which slides up and down the ladder, can be attached to the climber's belt and locks automatically if the climber should fall. It also includes a strut attached to the ladder which grips the stem and maintains a distance of about a meter between the bole and the top of the ladder. This enables the picker to reach further out into the crowns of small trees of species such as *Cupressus* which bear fruits near the ends of the branches.

General purpose single-section ladders extend up to about 11 m, while doublesection extending ladders extend to about 14 m. They may be leaned against a stout branch or against the bole for direct access to the crown. The top must be made fast to the tree with a nylon rope and a tall ladder must be further steadied with two guy-ropes. Depending on species and crown form, the climber may leave the ladder and climb into the crown, using safety belt, safety line and nylon safety ropes as from the tree bicycle, or he may collect fruits while standing on the ladder; in the latter case he should secure himself to the ladder by means of a nylon safety rope, while foot fatique can be reduced by a movable foot stand that can be hooked on to a convenient rung.

The larger and heavier ladders need a vehicle for transport between stands. Extension ladders mounted on trucks can be used to reach crowns of up to 20 – 25 m, but the lack of manoeuverability restricts the vehicle to roads or very flat ground. A light alloy ladder mounted on a tractor is described by Morandini; it will reach to 14 m and the tractor can be used in forests without roads. In the USA a magnesium extension ladder mounted on a halfton 4-wheel truck, can be used with safety to 9 m. Sway from the vehicle suspension is eliminated by the use of four mechanical jacks attached to the support posts of the ladder platform rack.

Ropes and hoisting equipment: Access to the crown can be achieved by suspending a rope, rope ladder or hoisting equipment from a stout branch. The same methods for projecting a thin line over the branch (throwing, catapult, shooting arrow) are used as when using a rope to shake branches. Because a stronger and heavier rope is needed to support the climber's weight than to shake branches, it is necessary to pull up the rope in three stages (1) Initially a light nylon line of 50 lbs. (23 kg) breaking strain which is used to pull up (2) A nylon cord of 3 – 4 mm diameter, which is used in turn to pull up (3) A nylon rope of 13 – 18 mm diameter, strong enough to carry the climber.

Rope ladders have been used with particularly successful results in Czechoslovakia. They may be up to 30 m long and weigh about 20 kg. Wooden rungs are set 30 cm apart. Hoisting equipment involves the use of a block and tackle which are hauled into position and secured by tying the rope firmly at the base of the tree. The collector is hoisted into the crown of the tree on a small bosun's chair or climbing saddle by one or two men on the ground or with the aid of a mechanical or electric winch. The method has an advantage over climbing irons or ladders because the ascent is less tiring and so reduces the risk of accidents brought on by fatique.

Tree net. Devices such as the rope ladder and hoisting equipment give access to the interior of the crown. Some genera such as *Cupressus, Chamaecyparis, Tsuga* and *Thuja* bear a large number of small cones near the tips of the branches, where they are not strong enough to bear the weight of a climber. To collect these it is necessary to reach the exterior of the crown. Mounted ladders provide one method of doing this. Another was the tree net.

The tree net was triangular in shape and measured 34 feet (10.3 m) across the base and 38 feet (11.5 m) from base to apex, with a 12 × 12 inch (30 × 30 cm) mesh. It was suspended by special ropes and snatchblocks from a point near the top of the crown, covering part of the exterior crown and with the lower corners pulled taut and attached by guylines to nearby trees or driven posts. The net carried two collectors at a time.

The net did no damage to the crown and was reasonably effective, once in position, on deep-crowned trees with an abundant cone crop but it was a very slow method at best and impossible to use in dense stands. Its use has been generally abandoned in favour of less time-consuming methods.

Climbing and Picking Fruits within the Crown

Methods of climbing and picking fruits in the crown are independent of the method used to reach the crown, whether this is by ladder, tree bicycle or climbing irons. The following summary of the main points to be observed is based on Yeatman and Nieman.

In climbing on the branches of a tree, as on the rungs of a ladder, handholds are used mainly for guidance and balance and the feet and legs are used for thrust. Only one limb - a hand or a foot - is moved at a time, gripping or standing on branches where they are strongest, close to the main stem. Doubtful branches should be tested by being pulled sharply before being trusted to carry the weight of the climber. Confidence and muscular coordination are keys to safe climbing. A sustained state of tension and reliance on strength alone can be both exhausting and hazardous. Most of the work should be borne by the legs.

Tension in a climber may be greatly reduced if he has confidence in a proven safety system. Occasional mishaps may occur even to the best climber, if only because a branch which appears safe turns out to be treacherously brittle or slippery. Modern climbing techniques and equipment are designed to ensure that, if a climber falls, he falls only a short distance before his fall is arrested.

Excellent illustrated accounts are available in Seal et al., Yeatman and Nieman and Robbins et al. and include details of the ropes and knots which the climber must know how to use.

Safety belt and strap: One essential item of equipment is the safety belt or harness. The most convenient type consists of a waist belt, which carries a number of metal rings to which may be affixed the safety line, safety strap or climbing chain, short lengths of safety rope, tool line etc., and a saddle which fits beneath the buttocks, is connected to the waist belt and can be used as a seat for the climber when this is convenient for collecting. Additional support can be provided by including a pair of shoulder straps as part of the harness. Attached to the safety belt are the equally essential safety strap or chain and one or more short nylon safety ropes which serve to secure the climber to the stem of the tree. One end is fastened to one side of the belt while the other is passed round the stem and then attached to the other side of the belt.

A light chain is sometimes used as the climber ascends the bole, the length of which can be readily adjusted to the taper of the tree by taking in a few links. An adjustable safety strap is as good or better. A second, reserve short rope should be carried to enable a climber using climbing irons or ladders to surmount an obstructing branch below the live crown; the reserve rope is fitted above the branch before the original one is unfastened below it. The safety rope is also fitted to the upper stem while the climber picks fruits within easy reach. Thus safety belt and safety ropes together will, on suitable trees, ensure the safety of a climber as he ascends the bole and while he picks fruits, even in the absence of a safety line. But they do not ensure safety as he climbs within the crown.

Safety line and safety ropes: Another important item is the safety line. Nylon rope of 12 – 14 mm diameter and weighing abour 1 kg per 10 m is commonly used. Apart from its strength and good wearing properties, it has advantage of being somewhat elastic, thus reducing the jarring effect on a climber's body as the rope tautens after a fall. Polypropylene is less suitable since it melts at relatively low temperatures. However, it has been used successfully in some countries e.g. Honduras, where nylon is inobtainable. Precautions taken are to use polypropylene rope at least 18 mm in diameter and to avoid any techniques which would involve prolonged friction on a stationary rope. Temporary friction of a moving rope, e.g. over a branch, is not dangerous since the heat generated at any one point in the rope is very little. Natural fibres such as hemp or jute need to be substantially heavier than nylon if they are to attain the same strength, in addition they are subject to rot especially if used in wet weather.

Safety lines or ropes may be conveniently divided into two classes, the short safety rope which the climber uses to attach himself to the tree while working in the crown and the long safety line which runs down to the ground and is controlled by an anchorman. The short safety rope varies in length from 3 to 10 m and is of the same gauge as the long safety line; a climber will normally carry two of them attached to his safety belt. When he is working in the crown, he attaches one of them either to a stout branch or to the stem. If he is in the upper crown where the stem is too thin to be safe, he attaches the rope to the stouter portion of the stem *below* him and passes it spirally upwards round the stem. If the stem is stout enough and he uses a safety rope of appropriate length, he can reach well out into the outer crown. The short safety rope is simple to use and renders the climber completely independent while collecting in the crown.

One end of the long safety line is attached to the climber's safety harness, the other is controlled by an anchorman who remains on the ground. The length of the line must be at least twice the height of the tree, to allow a climber to be lowered to the ground, if necessary, from any position close to the top of the tree where he may be working. Apart from the additional safety factor it provides, the safety line can enable the climber to reach fruits in the outer crown which would otherwise be inaccessible.

Method of climbing with use of safety line: On reaching the lower crown, the climber ascends in a slight spiral round the stem, so that he takes his long safety line across branches which will hold the line in the event of a fall. If live branches are lacking on one side of the stem, the climber should follow a zig-zag course on the side of the stem bearing live branches. If stout branches are lacking, the climber should fit a safety rope to the stem and clip the safety line to it. The aim should be to ensure that he cannot fall freely more than 2 m before the safety line starts to take his weight.

Meanwhile the anchorman, as described by Yeatman and Nieman, must stand well clear of the fall line from the tree being climbed to avoid possible injury from falling branches that may be broken off as the climber ascends. The anchorman must be ready at any time to support the weight of the climber on the safety line as he feeds it under one arm, around his back, and over the other shoulder, releasing it alternately with one hand while gripping it with the other. A half turn of the rope should be made around the lower trunk of a neighbouring tree to provide additional friction and greater security should the climber slip or fall.

However, the friction should be minimal during the ascent to ensure that the climber is not held back by the safety line. It is important that the unused part of the safety line remains coiled on the ground and free of obstruction so that tangles are avoided as the rope is paid out. The long safety line is of particular advantage when it is necessary to take rapid evasive action from wasps or bees that are sometimes encountered in tree climbing.

When the climber reaches the level in the crown where he is to collect, he may secure himself by passing the safety line over a stout branch and around the stem. Alternatively he may secure himself to the stem by means of a safety rope passing it over a stout branch and around the stem above him and re-attaching it to his safety belt then detaching the safety line. If the branches are not stout enough, he fixes a short safety rope to the stem as high above him as possible and passes the safety line through it. The rope should never be attached to the stem above the point where the diameter falls below 8 – 10 cm.

The climber can now either go on to pick the fruits with the anchorman still holding the safety-line or he can apply a special safety line locking method, using carabiners or safety clips, and take charge of his own movements. In either case most of his weight can be supported by the harness saddle, enabling him to reach further out in the crown than would be the case if his entire weight had to be supported on relatively thin branches. Use of the safety line locking method allows the climber to make a controlled descent as he works downwards collecting fruits as he goes.

An advantage of this method is that the anchorman is freed temporarily for other work; in this way one anchorman can look after several climbers. It is well suited to species in which cones or fruits are scattered all over the crown and picking in one tree takes some time. In species in which fruits are clustered in one part of the crown, picking is quick and it is best to keep the anchorman on the line.

Method of descent: For descent the climber must first climb up and release the topmost safety rope if one was used, while the anchorman takes in the slack of the safety line. Then, as described by Yeatman and Nieman, the climber descends with care, following the same route down as the one he ascended by, so as not to foul his safety line on the small branches of the upper crown. When his feet reach branches that will support all his weight on the safety line and are large enough in diameter to withstand rope abrasion, he moves laterally to catch the safety line on the crotch of a branch.

The climber continues down so that the trailing safety line passes up and around the stem, over a stout branch, and on around and down to the climber. Alternatively, in trees with dense crowns, it is often easier to stop, attach oneself to the tree, untie the safety line, pass it up on one side of the trunk over a branch on the opposite side and down the other side, and retie it to the body belt. With the safety line securely caught on a branch (crotched), the climber may be let down by the anchorman paying out the rope or he may control his own descent by using the safety line locking method.

Descent to the ground by safety line is the best method when the ascent of the bole has been by climbing irons. When ascent has been by tree bicycle or sectional ladder it is usually preferable to descend the some way; otherwise the climber must loosen the bands of the tree bicycle or disconnect the sections of ladder while suspended from the safety line.

Method of picking fruits: With correct use of the safety harness and safety line the climber should have both hands free to pick the fruits. Methods vary according to the size, number and distribution of fruits and the firmness of the peduncular attachment. Numerous, small, clustered and accessible fruits can be picked and deposited immediately in a bag attached to the climber's belt or slung from his shoulder and with its mouth braced open. Bigger and more scattered fruits may be detached, dropped and collected from the ground later. Fruits too inaccessible to be picked by hand may be detached by long-handled pole, hook, rake or shears. Tools used in the crown should be pulled up by the climber by the use of a light tool line attached to his belt, after he has reached the collecting area in the crown.

In Chile climbers beat the branches with a long stick to detach fruits of *Nothofagus alpina*, but the method is less effective with *N. pumilio*. Some form of hook is in common use for conifers. In Thailand the hooks used for collecting cones of *Pinus kesiya* and *P. merkusii* have been designed with two sharpened edges pointing in opposite directions, so that they can be used to detach cones with either a pushing or a pulling action, thus avoiding the breakage of twigs with new conelets. They are mounted on light steel tubes which can be extended to a length of about 3 m. Various types of hook, rake and shears are illustrated by Robbins et al. Gathering of fruits from the ground, when they have fallen after detachment by tools, is the same as described above for natural seed fall.

If fruits are small, numerous, clustered, inaccessible for hand picking or held firmly by the peduncle, it may be necessary to cut off whole fruit-bearing branchlets by long-handled shears or saw. *Pinus oocarpa* is an example of a species in which the cones have strong woody peduncles which are difficult to sever without damaging the branchlet. In such cases one alternative is to collect by cutting off the branchlets, but this causes destruction of the next year's crop of conelets. A better solution is to devise new equipment to overcome special local problems. In Honduras a bell-shaped sharpened hook is designed to be pushed outwards from the centre of the crown towards the ends of the branches and this severs the cones while doing the minimum damage through bending or breaking the branches. The hook is mounted on a 5 m aluminium or wooden handle. It has been in use for over 5 years and has proved to be very practical for cone harvesting from *P. oocarpa*, *P. caribaea*, *P. patula* subsp. *tecunumanii* and *P. maximinoi*.

If fruits are clustered at the end of long branches where they are out of reach of the climber, he may need to cut off the whole branch. In Brazil climbers cut off fruit bearing branches of *Eucalyptus* with a knife on a pole and the usual practice is to collect from alternate trees in successive years so that the tree crowns have a chance to recover.

Climbers should wear boots with non-slip soles and overalls without belts or loops which could catch on branch snags. A close-fitting safety helmet, gloves and safety glasses are often useful. Anchormen must wear industrial safety helmets to protect them from objects falling from the trees. A light pruning saw is often necessary to provide a clean bole suitable for climbing by tree bicycle.

The methods described above have proved highly reliable when used on coniferous species. There is less experience and probably greater difficulty in climbing hardwood species of the tropical moist forests, where a few massive branches separated by intervals of 3 – 4 m pose a special problem. Very thorny species such as *Acacia* spp. present another problem but, because they rarely exceed 15 – 20 m in height, collection can often be made from the ground.

Useful aids to collection described by Doran et al. include a strong vehicle roof carrier to stand on, light ladders fitting on the roof carrier, a rope and weight for throwing over a branch to shake or break it, flexible saw or long-handled pruning shears and thick leather gloves for stripping pods from thorny species. Standard collection methods will certainly need to be modified to adapt them to the peculiarities of individual tropical species.

Collection from Standing Trees with Other Means of Access

Some types of equipment have been designed to raise the collector mechanically to a level where he can reach the fruit-bearing portion of the crown, without having to climb at all. Limited trials have shown that cable systems supporting a carriage can move pickers alongside tree crowns. Although access to several trees is obtained with one setting, the installation is time consuming. The system would prove most profitable in a stand where repeated collections are to be made. Experiments have also been made with cables or platforms suspended from balloons or helicopters. Balloons are not considered practical, but helicopters may have a part to play in certain conditions.

Probably the most practical of all mechanical devices to raise or lower a man to within reach of the fruit crop is the Extension Platform. The following account follows that of Turnbull. The extension platform of a type used for the installation of overhead electric cables has been used for seed collection in many countries. There are a number of models available, including ones with telescopic raising gear and hydraulically operated articulated steel booms built on a turn-table.

Matusz describes the AGP-12 articulated extension device which has been constructed in the USSR and Seal et al. illustrate and explain the use of the British developed Simon hydraulic platform. The Russian machine could be used to collect seeds from trees to a height of 15 m and the Simon hydraulic platform to a height of 10 – 16 metres. They provide adequate working space for two men, who can dispense with safety-belts and associated gear. It was intended to develop a larger version of the Russian machine to reach to 30 m.

A trailer-mounted platform attached to, and powered by, an agricultural tractor has been developed in Australia for seed collections from heights up to 10 metres above the ground. This tractor-trailer unit is capable of being used almost anywhere a tractor can be safely worked. It is versatile and relatively low cost equipment. The Afron Hydraulic Drive Power Ladder has a lower reach (maximum picking height 7 m), but is highly manoeuvrable and is operated by a single person, since all the controls are on the picking platform. It is widely used in fruit orchards and could be a useful machine in intensively managed forest seed orchards provided that the trees are of moderate size.

Figure 3. Use of hand tools and safety line in picking fruits in the crown.

According to Seal et al., the hydraulic platforms are of most value when time and labour are short and fruits can be collected from accessible trees with good crops.

Disadvantages of this equipment are the necessity for good access and the high capital cost. Where cone crops have been less accessible and prolific, the per unit cost of collection using the Simon hydraulic platform in the UK was found to be greater than using ladders or tree bicycles".

Productivity in Fruit Collection

The productivity in fruit collection increased especially for tropical species. As pointed out by Dobbs et al., amounts collected per picker-day depend on a variety of factors which include not only the skill and energy of the picker but the size of fruits, the heaviness of the crop, the firmness of attachment of the peduncles, the presence of old or immature fruits on the branches which may confuse or slow down the pickers, the method of collection (by climbing, from felled trees, from the ground etc.) and miscellaneous factors such as weather, insects and travelling time.

For *Pseudotsuga* in USA and Canada, Douglass and Dobbs et al. quote the same average figure of 2 – 3 hectolitres of cones per picker-day, collected from about four trees by climbing. Dobbs et al. indicated that the same rate of production can be expected from felled *Pseudotsuga* trees as by climbing the standing trees.

4 – 5 hectolitres of cones per climber per day can be expected for *Pinus ponderosa* and around 0.5 hectolitre for smaller-coned species such as *Larix, Thuja* and *Tsuga.* In IUFRO collections made in *Pseudotsuga, Picea sitchensis* and *Abies grandis* in western North America, a team of 4 experienced climbers could complete one selected stand in a day, collecting from about 20 well spaced trees.

Figure 4. Climber controlling own descent by safety line locking method.

In Thailand in stands of *Pinus kesiya* of medium size trees an average of 25 – 30 kg of cones is expected per climber-day, collected from 6 trees. In Honduras the collecting rate for *P. oocarpa* and *P. caribaea* averages 1 to 2.5 hectolitres from 3–5 trees per man-day, for a general bulk collection in an average crop. Matusz reported that an average collecting rate in European pine stands, 25 m tall and bearing average to good crops, with the use of climbing irons, was 20 – 50 kg of cones from 8 – 10 trees in a day, yielding 0.4 to 1.0 kg of seeds. For each tree the climber spends 15 – 25 minutes in ascending and descending and 20 – 30 minutes in picking cones. In Denmark the following yields of fruits or cones have been achieved, given the conditions of:

1. Easy access to stands and the standing trees.
2. A very good seed crop.
3. A very experienced team of climbers, using ladders and working 8 1/2 hours a day.

Species	*hl fruits or cones/man-day*		
	good	*max.*	
Abies nobilis	10	14	
Abies Nordmanniana	3	5	
Abies alba	5	7	
Acer pseudoplatanus	25	35	
Fraxinus excelsior	30	40	
Picea sitchensis	2	3	
Picea abies	4	6	
Picea omorica	1	1.5	(young stands)
Larix leptolepis	1.5	2	(young stands)
Larix decidua	0.6	1	
Pinus sylvestris	1.8	2.5	(seed orchards)
Pinus austriaca	1.8	2.5	
Pseudotsuga sp.	1.5	2.0	

Yields of seed per ha in a very good seed year are:

Abies alba	150 – 200 kg
Abies Nordmanniana	150 – 200 kg
Abies nobilis (procera)	250 – 600 kg
Larix sp.	25 – 70 kg
Picea abies	50 – 150 kg

Picea sitchensis	15 – 25 kg
Pseudotsuga sp.	40 – 80 kg
Fagus sylvatica	500 – 1000 kg
Quercus sp.	2000 – 5000 kg (?)
Fraxinus excelsior	400 –800 kg
Acer pseudoplatanus	200 –400 kg

Little information is available for fruit collecting rates for tropical hardwoods.

Training and Safety

Seed collection, especially by climbing, is arduous work and it is essential that climbers are carefully selected and well trained before they commence collecting operations. They need to be physically and mentally fit, with a natural aptitude for climbing and a combination of self-confidence and common sense. Any sizable collection programme should have a nucleus of at least one skilled climber on the permanent staff, who may be employed on other duties outside the seeding season. It will be his responsibility to conduct short training courses for any temporary climbers before the start of each collection season. Good pictorial illustrations are an invaluable training aid, especially where climbers are illiterate.

Safety precautions will vary according to local conditions and particularly the species of tree and the equipment and methods of collection used. All staff taking part in collecting operations should be fully conversant with local safety rules. The selection of safety hints reproduced below:

1. All equipment should be carefully stowed, both during transport in the field and while in store between collecting seasons.

2. Clothing should be strong, well fitting, and suited to the weather expected.
3. All equipment should be checked before it is used and, if there is doubt about its condition, it must not be used until repaired or replaced.
4. Do not climb in wet or very windy weather, nor in poor light as at dusk, nor when overtired.
5. Do not climb trees with obvious signs of stem rot, severe cankers or galls, split stems, double leaders, or other abnormalities indicative of mechanical weakness.
6. The safety line should be coiled on the ground before the climber ascends to avoid tangling or snagging the rope in the underbrush.
7. The anchorman should hold the safety line under one arm and over the other shoulder. It is wise to make a half turn around a neighbouring tree. This gives control and prevents the safety line from being pulled from his hands. Pull in and pay out the safety line by alternate hand grips. A sliding rope is difficult to control and can cause painful friction burns.
8. Never climb with anything tied or looped around the neck.
9. Safety helmets and goggles should be worn to prevent injury to the head and eyes in climbing rough, densely branched trees.
10. Stand on and grip branches close to the point of attachment to the main stem.
11. Watch for brittle branches; test doubtful branches before putting weight on them. Avoid branches with bark peeling from them - they are slippery. As

far as possible, decide on the climbing route while still on the ground, especially for the branchy crown region.

12. The climber should have three points of support at all times (one hand and two feet or two hands and one foot), moving one limb at a time, except when attached to the tree by a safety strap or rope or when suspended on a safety line. Climb calmly with regular movements, taking short steps.
13. Do not carry tools while climbing the crown. If there is need for a pole pruner or cone rake etc., use a light tool line to hoist the equipment to the working level. Leave the tool line attached to large tools as a lanyard while working. Return tools to the ground on the line, do not drop them or throw them down.
14. Beware of sharp branch stubs: they can snag clothing and may cause painful cuts and bruises.
15. Climb spirally or in a zigzag manner, or fasten safety strops to the stem so that you cannot fall more than 2 m before your weight comes onto the safety line.
16. The diameter of the main stem should not be less than 8 cm at waist level during climbing. If in doubt concerning security, do not hesitate to tie a safety strop to the stem at a safe level before climbing within reach of the seed-bearing crown.
17. While attaching safety rope, keep one arm securely around tree until the rope is fastened to safety belt.
18. Before letting go of the tree with your hands, test your weight against the safety rope and footholds.
19. When picking near the top of a tree, keep your body close to the stem, so that your weight bears down, not outward.

20. The safety strap should always be attached around the tree stem except while you are climbing or changing position in the crown or are suspended on the safety line.
21. Before dropping bags of cones or other material, be sure that the personnel on the ground are notified and are well clear.
22. When collecting fruits from a ladder, make fast the top of the ladder to the tree with a nylon strop. The ladder must be further steadied with two guylines.
23. Have a well-stocked first aid kit handy at the climbing site at all times.

3

Planning for Seed Production

In the majority of species the seeding season is concentrated within a few weeks and the collector's objective is to collect as much of the crop as possible within the short period while the seeds are mature but the fruits have not yet fallen or dehisced. Large indehiscent or fleshy fruits may be collected from the ground but even in these cases collection must be done quickly to avoid losses from animals, fungi or premature germination. Prior planning of collecting activities is therefore essential in order to ensure that operations are conducted as quickly and efficiently as possible in the limited time available.

Collection in accessible and easily observable plantations or seed orchards reduces the need for careful preparation. On the other hand, collection in inaccessible, multispecific natural forests or sampling a number of different seed sources within a widely spread species calls for very careful planning if trained collecting teams are to operate with the right equipment in the right place at the right time. International seed expeditions encounter special problems, as they often operate in several different countries, each with its own regulations, and they seek to supply the varying needs of many using countries.

Selecting Species, Provenances and Stands

Species

Selection of species for planting often presents no problem. In a simple afforestation project, which uses a proven well adapted species and provenance and obtains the seed from a local seed source, the choice is automatic. But not infrequently afforestation objectives change, e.g. emphasis may shift from sawlog to pulpwood or fuelwood production, or unexpected disease problems may arise. In East Africa *Pinus radiata* featured prominently in planting programmes until it was severely attacked by the needle blight *Dothistroma pini* in the 1960's. Thereafter its large-scale planting had to be abandoned and the planting programmes of resistant species, such as *P. patula* and *Cupressus lusitanica*, were expanded in compensation.

For large-scale collections, data on seed demands by species need to be assembled some months in advance. Most species need a year or more in the nursery. Estimates of seed demand must therefore be made about two years before planting in the field. Only rarely will the seed collectors and the seed users be the same persons. More often, whether collection is done by forest services or by private collectors, they will be supplying the needs of several different users. A centralised organisation is needed to solicit demand estimates from the several planting agencies and to consolidate these by species and provenance. Consolidated regional or global estimates of seed demand are much more difficult to compile than those for single countries, but recently attempts have been made to do this for tropical conifers and for western North American conifers.

Provenances

The word "provenance" has been used in somewhat

different ways by different authors. In its simplest use, it is "The place in which any stand of trees is growing". When applied to seeds, the meaning is frequently extended to include "The area where the mother trees of the seeds were growing". In cases where seeds are collected from an exotic plantation or "derived provenance", there has been some inconsistency in usage; some authors would define provenance as the place in which the immediate parents were growing as exotics, others would confine its use to the place where the original progenitors were growing in natural forest.

Provided that seed origin data give the data on the full pedigree, including both the location of the original natural progenitors, the location of the immediate parents and the location of any intermediate generations, e.g. Cmpt. K2 Elburgon, Kenya (immediate parents) ex Cmpt. 16 Nelspruit, South Africa ex Los Reyes, Hidalgo, Mexico (original natural forest progenitors), the practical forester will not worry as to which is, strictly, a provenance and which is not.

Over the last half century evidence has continually accumulated that, within a botanical species, significant genetic variation in forest trees is frequently associated with geographic differences between the places where they are growing. This is particularly the case where geographic displacement is associated with climatic or soil changes. Thus the word "provenance" is increasingly applied to areas characterised by the genetic nature of the populations growing there rather than their location alone, e.g. "The geographic source or location to which plants are native and within which their genetic characteristics have been developed through natural selection". For the purpose of seed collection, the ideal provenance, as described by Barner, would:

a) Be composed of a community of potentially interbreeding trees of similar genetic constitution (and of significantly different genetic constitution from other provenances).
b) Be sufficiently large for collection of reproductive material in quantities significant for forest practice.
c) Be defined by means of boundaries which can be identified in the field.

Although it is not yet possible, in most cases, to delineate the boundaries of natural provenances, there is ample evidence in many tropical as well as temperate species that significant genetic differences exist between them. In the case of derived provenances growing in plantations, boundaries can be defined much more easily and, after a generation or two of conscious selection by man, these "land races" often differ significantly from the original natural provenance.

Increasingly foresters recognise the vital importance of provenance and specify the precise provenance which they need to plant on a given site, not just the species. Even within a single country distinct provenances or races of a species are recognised; some may be morphologically distinct, others which look alike may differ in their adaptability to specific sites. A good example is the large number of provenances of *Tectona grandis* recognised in India. Seed collecting teams must therefore expect an increasing number of orders with a breakdown by provenance as well as species. This trend is to be encouraged but it does complicate collecting operations, since it is clearly more time-consuming to collect say 20 kg of seed from each of 10 different locations 100 km apart than to collect 200 kg in one area.

Another difficulty is in deciding on the limits of a provenance. Often a provenance is named from the nearest village and there is no evidence to show whether there is a significant shift in gene frequencies in the populations 1 km, 10 km or 100 km away from the original collecting point. Some attempts have been made to define the limits of provenances or seed zones of a few north-temperate conifers and similar studies have been made more recently on the delineation of provenance regions of *Pinus caribaea* and *P. oocarpa* in Honduras.

Provenance regions of *Eucalyptus camaldulensis* have been defined in terms of the main drainage systems in Australia but very little work has been done on this aspect in the tropical hardwoods. The need to collect more than one provenance of a species calls for increased care in planning field operations. Some widespread species flower and fruit a few weeks earlier at low than at high elevations and at lower than at higher latitudes. Knowledge of a species' phenological variation in relation to geography will assist the collector to choose the most appropriate sequence of collecting sites so as to extend the total duration of useful collections.

Stands

In contrast with provenances, the boundaries of individual stands are commonly well defined. In many cases the stands are being managed for seed production e.g. by thinning. Often they are in plantations. Seed orchards are a special case, designed for seed production before they are planted and managed continuously for that purpose. The problem with seed stands and seed orchards is thus not one of identification, but that their area may be insufficient to supply all the demands for seed placed on them. If this is likely to occur, it is advisable to require seed users to name second or third choices, in both stands and provenances, in case their full needs cannot be met from their first choice.

Verifying Seed Quantities

Seed users need to define the quantity of seed needed of each species, provenance or stand. For this it is necessary to know the area of plantation to be established annually and the initial spacing to be used, together with an estimate of losses and culls in the nursery, of replacements needed after planting to achieve full stocking, and of the number of germinated seedlings to be expected from each kg of seed sown. An example of the type of calculation is shown in Table 1.

Information on planting area and initial spacing is usually available from Plantation Management Plans, while some guidance to germination rates is available in published documents. Whenever possible, local experience on variation between provenances and planting sites should be used to refine estimates based on average conditions. For example, seeds of two provenances of *Picea abies* weigh, respectively, 6 gm and 12 gm per 1000; seed of *Eucalyptus cloeziana* collected in the moist coastal forests of Queensland averages 100,000 – 400,000/kg, whereas seed from dry inland wood lands averages only 35,000 – 65,000/kg. In Italy it was found that in nursery trials on several eucalypt species the number of plants produced as a percentage of viable seed ranged from 18 % for *E. robusta* to 46 % for *E. camaldulensis*.

Similarly, differences in climate, soil and incidence of pests and diseases can have a big effect on the rate of losses in different nurseries and plantations, whether or not there are any differences in the efficiency of management. So it may be necessary to apply an appropriate "locality correction factor" or "nursery recovery factor" to arrive at an accurate estimate of seed requirements for a particular plantation project. Before sending in his final order for seed to a central seed unit or commercial seed merchant, the plantation project manager should deduct those quantities of seed already in stock or likely to be available by collection from older plantations within the project area.

Seed content is estimated by counting good seeds on one surface of each of several sliced cones.

An alternative approach, favoured in some countries, is for the project manager to specify the number of plantable plants he needs to raise and to leave it to the seed officer to decide on the weight of seed to be collected and issued, in the light of germination tests of current seed lots and of known nursery recovery factors. If good annual crops of seed can be guaranteed, it is satisfactory to order enough seed each year to produce nursery stock for the area scheduled for planting in about two years' time. Storage space can then be kept to a minimum.

For species which show periodicity in seeding, however, there are big advantages in acquiring several years' supply in a single good seed year, when the seed will be both cheaper and better. This course of action can only be of practical value if local seed storage facilities are adequate to maintain seed viability over the interval between good seed years.

Table 1. Estimating Seed Demand

1. Species	(a) *Pinus kesiya*	(b) *Tectona grandis*
2. Plants per ha		
(a) Number planted	1670 (3 × 2 m)	1111 (3 × 3 m)
(b) Additional for replacing field casualties		
- percent	15	35
- number of plants	250	389
(c) Total requirement - plantable plants	1920	1500
(d) Additional for nursery losses and culls		
- percent	20	62.5
- number of plants	480	2500
(e) Total requirement - germinated seedlings	2400	4000
3. Estimated number of germinated seedlings per kg of seed received	32000	500
4. Basis for estimate in 3.	1980 crop ex Zambia, dewinged and cleanel	1979 crop ex Trinidad, with involucres removed
5. Number of kg of seed needed per ha of plantation	0.07 (13.3 ha/kg)	8.0 (0.12 ha/kg)
6. Annual planting area (ha)	12000	5000
7. Annual requirement of seed (kg)	900	40000

Knowledge is needed for each species or provenance on both the likely interval between good seed years and the rate of loss of viability of the seed under existing storage conditions. If heavy seed years are separated by several successive years of seed crop failure and if storage at ambient temperature results in rapid loss of viability, the alternatives may be to build a refrigerated storage room or to change the species.

Determining the Year

Effect of Periodicity

Seed-bearing of many forest trees is rather irregular from year to year. One year with a heavy crop (a "seed year" or "mast year") may be followed by one or several years with a poor seed crop or none at all. This habit of periodicity in seeding is an important factor to consider when planning seed collecting operations. The collection of seeds in a good year confers a number of advantages. There can be a high intensity of selection of seed bearers, the cost of collection is lower, due to the concentration of the crop, and the seeds will usually be of higher germinative capacity and will retain their viability longer than those collected in a poor seed year. Damage from insects affects a smaller proportion of seeds in a good than in a bad seed year. A heavy seed crop usually reflects a previous heavy production of pollen, to which all or most trees in the stand have contributed. Collection in a good seed year therefore conserves a higher proportion of the genetic diversity among male parents than collection in a bad year which follows pollination from only a small number of trees.

Periodicity is well documented for many temperate conifers. For example on average *Pinus sylvestris* bears an abundant crop every 2 – 3 years and *Pseudotsuga menziesii* every 4 – 6 years in the UK. Since the period between the good crops is not regular, a general rule of collecting three years' sowing requirements whenever a species bears a heavy cone crop is there recommended by Seal et al.

Periodicity in tropical species is less well documented. Irregular mast years in *Triplochiton* have a great influence on the regeneration, or lack of regeneration, of that species, although periodicity in pests and diseases (the weevil *Apion* and the fungal smut *Mycosyrinx*) may play as big a part in seed production as periodicity in flowering. Poor seed years have been recorded in *Pinus caribaea* and *P. oocarpa*, and *P. merkusii*. In other species periodicity is not marked. *Tectona grandis* has generally good flowering each year, although exceptionally good seed years are observed in some localities on a three or four year cycle.

Gmelina arborea starts to seed early, from age 3 in the Philippines to age 7 in Nigeria and usually produces regular abundant crops but poor seed years have been recorded, at least in some provenances. *Pinus kesiya* bears abundant crops every year within its indigenous range and as an exotic if planted in the appropriate climate. *Cassia siamea, Acacia mearnsii, Cupressus lusitanica* and ornamentals such as *Delonix regia* and *Jacaranda mimosaefolia* are other species which can be expected to flower and fruit profusely each year.

Periodicity may vary considerably between species within the same genus. Among eucalypts, *E. grandis, E. saligna* and *E. camaldulensis* usually bear heavy seed crops every two to three years, while *E. gomphocephala* and *E. maculata* only seed heavily at longer intervals. Dipterocarps in Malaysia have heavy seed years at unpredictable intervals of one to six years. Periodicity and flowering patterns of eucalypts can change when they are grown as exotics. *Eucalyptus maculata* and *E. citriodora* bear much larger crops more regularly when grown in plantations.

Even in good years, flowering may vary substantially from one locality to another. Sometimes individual trees of a stand are on different cycles, some flowering abundantly one year, others in the next.

Counting the Fruit Crop

For those species which are known to exhibit periodicity in flowering and fruiting, it is highly desirable to visit the stands to be collected well in advance of the fruiting season, in order to assess in which of them the next seed crop promises to be heavy enough to justify the cost of collection. Too little is known about which external factors have a decisive affect on flowering to allow prediction of future seed crops on the basis of climate. Estimation of the crop can best be done by counting flowers or young fruits on a sample of the trees in the collection stands. Assessment of the abundance of flowering can give a preliminary estimate of the potential seed crop, but may be misleading if there are subsequent severe losses e.g. from insects, wind or poor pollination.

In *Eucalyptus regnans* trapping studies indicated that only about 15 % of flower buds and 30 % of flowers can be expected to develop into mature fruits. In species such as the pines, which take two years from pollination to ripen their fruits, a count of one year old cones can give a useful indication of the next year's crop, which can be confirmed by further inspection a month or two before collection is due to begin. Recurrent personal inspection of the future crop is the ideal and presents no problem if collection is carried out in seed orchards or easily accessible plantations or natural forest. It may be difficult or impossible for collecting teams working in inaccessible areas or for international expeditions working in several countries. In such cases the team leader may have to rely on reports from an experienced correspondent or on estimates made during the previous year's collections. If reliable local information is not available, a special reconnaissance may be justified in advance of the expensive collecting expedition.

Where the main seed collecting areas are located in less accessible places, it is useful to maintain permanent phenological plots in easily accessible locations within the same forest type, to act as biological indicators. Such plots should be monitored regularly so as to obtain a record of the timing and intensity of flowering and fruiting. The performance of the plots will indicate to collectors the best times to go into the less accessible areas of the district, to check for flowering. The boundaries of a phenological district that can be usefully served by a phenological plot must be defined by experience.

In Malaysia, a sample of 86 Dipterocarp trees in an artificially established arboretum at the Forest Research Institute, Kepong, is monitored monthly or fortnightly and the percentage of trees flowering in a month or year is used as an index of Dipterocarp flowering for that month or year. This index gives a fairly reliable indication of Dipterocarp phenology within the State of Selangor (about 8000 sq. km) in which Kepong is located. Since the phenological plot in Kepong is within 10 minutes' walking distance from the researchers' offices and laboratories, a lot of time and money is saved which would otherwise have to be spent on transportation and organisation of field trips.

Where it is necessary to count the fruits or cones, binoculars or telescopes are an essential aid. They should be of high optical quality. In binoculars a wide field of view should be combined with only moderate magnification; a minimum of 50 mm aperture and magnification x7 or x8 are suitable. The normal method is to make counts on a representative sample of seed trees dispersed throughout the whole seed source. It is necessary to take the sample from within the stand, because the perimeter trees always fruit more heavily than the trees within the seed source.

Counting may be done from the ground or by climbing neighbouring trees. Fruits are counted from one side of the crown only and the number counted is converted to an estimate of the total crop for that sample tree by a correction factor which varies according to the species and the abundance of the crop. In Czechoslovakia (counting from climbed trees) a factor 1.6 is used for *Abies alba,* which is constant irrespective of the size of crop, because the clustering of the cones near the top of the tree makes counting in this species reliable. For *Picea abies* the factor varies according to the average number of cones counted per tree; for 1 – 40 cones it is 1.4, for 41 – 70 cones it is 1.8 and for more than 70 cones it is 2.5. In the UK (counting from the ground) a factor of 4 is used for *Pinus, Larix* and *Pseudotsuga,* while for species with very numerous small cones the practice is to scan say one tenth of the crown from one side and use a factor of 20.

The number of sample trees used for cone counting varies according to the size of the stand. In the UK five trees are sampled in small seed sources less than 0.5 ha and the number increases progressively to 20 trees for seed sources of more than 4 ha. Czechoslovakia uses a series of plots, in each of which about 5 dominant trees are climbed to scan a further 10 – 15 neighbouring trees. In Tasmania extensive eucalypt seed sources are sampled at the rate of 1 tree per ha.

Fruit crop rating methods

The results of cone or fruit counting are applied to the stand and expressed as a numerical value on a scale running from total crop failure to exceptional seed years. With experience it may be possible to define the criteria for an economically collectable crop in quantitative terms e.g. for *Pinus sylvestris* in the UK a minimum of 25 seed trees per ha each bearing at least 300 – 400 cones is specified. For conifers in Arizona and New Mexico Schubert and Pitcher defined "Few" cones per tree as 1 – 20, "Many" as 21 – 160 and "Loaded" as over 160 cones per tree. Quantitative classifications of this type will clearly vary to a great extent according to species, provenance and site conditions.

Rating	*Explanation*
Other than true firs:	
5 Heavy	Good crop of cones on all exposed crowns of most trees.
4 Medium	Good to medium crop on ¾ of exposed crown on most trees.
3 Light	Good to fair crop on ½ of exposed crown of ½ of trees
2 Very light	Some cones on some trees.
1 Failure	No cones to a few scattered on a few trees.
True firs: (upper 1/3 of crown)	
1 Failure	No cones to a few scattered on a few trees.
5 Heavy	Good crop of cones on most upper branches of most trees.
4 Medium	Good to medium crop on most trees.
3 Light	Few cones on many trees.
2 Very light	Few cones on scattered trees.
1 Failure	Nonexistent

More often use is made of qualitative scoring assessments which rely on the experience of the assessor. In Washington and Oregon five ratings are used, as shown below:

Each year the State Forest Service publishes estimated average cone crop ratings by species and geographic area, for the benefit of individual pickers. They are based on surveys of a number of different stands and the average for each area is expressed to one place of decimals. For example in 1972, a poor year, the best rating was 2.5 for *Tsuga heterophylla* in the Western Cascade of Oregon, while 1.0 or total failure was recorded for some species in more than one area.

In Tanzania a four class scale is used and seed crop estimates are made twice, one at flowering and the other about a month before seed collection. The classes are:

1. *no seed crop*. Trees without flowers and fruits.
2. *weak seed crop*. Flowering and medium size seed crop on free growing trees and trees on free borders of stands.
3. *medium size seed crop*. Flowering and very good crop on free growing trees and on free borders of stands, trees within the stands bearing crop at the top of crowns.
4. *very good seed crop*. Flowering and very good crop on most trees.

In Sweden forecasts of the cone and seed germinability of *Pinus sylvestris* and *Picea abies* have been made annually for about 80 years. Estimates are given separately for the different combinations of latitude (in 1° steps) and altitude (in 100 m steps).

Estimating full Seed Content by Cutting Test

The methods described above provide an estimate of the cone or fruit crop. It is necessary to relate this to seed production by examining the contents of a sample of the fruit crop. Fruits may develop normally to maturity whether one or one hundred of the contained ovules have been successfully fertilised and undergone normal development; in parthenocarpous species fruits can mature without containing any sound seeds at all. The number of fruits is therefore not always a good guide to the number of seeds.

The method generally recommended is to cut cones or fruits lengthwise and count the number of seeds which can be seen on one cut surface. Special cone-cutting knives have been designed for this purpose. One or two cone samples from each of 20 to 100 trees in an area are suggested for southern pines in the USA, 5 or 10 cones from each of 10 trees in the UK. Only normal seeds should be counted. Underdeveloped seeds which often occur at the top and base of cones are not included. The number of full seeds counted which denote a good crop varies according to species, for example 6 or more indicates a good crop in *Pseudotsuga*, 14 or more a good crop in *Picea sitchensis*. The relationship of the total number of full seeds per cone to number of full seeds exposed on a cut surface is known for some species, for example a factor of x4 or x5 is appropriate for *Pseudotsuga* in the western USA.

The average number of seeds per fruit for many tropical species is not known and needs to be established under local conditions; they vary from one seed per fruit in e.g. most Dipterocarps to several hundred per fruit in *Anthocephalus*. With multiseeded fruits, it is likely that the number of seeds developed will vary in accordance with climate, soil fertility and the age of the parent trees. The first crops borne by young trees nearly always contain fewer sound seeds per fruit than those borne by the same trees when fully mature.

Examination of a sample of seeds in the fruit serves the additional purposes of indicating the state of development or maturity of the seeds and the incidence of damage by pests or diseases. A final decision on whether the seed crop is sufficiently heavy to justify collection in a particular year should be dependent on *both* the rating of the fruit or cone crop *and* the results of the cutting test for determination of full seed content.

Dates for Collection

Some species in the tropics carry some ripe seed at all times of the year. Even in these there is often a period of maximum seed production, when collection will be cheapest and seed quality highest. In other species, and especially in the temperate zone with its marked distinction between summer and winter, ripe seed is borne for a limited period, often during the autumn. For many species there is good information on average dates of the seeding season, but these averages may not be sufficiently accurate for planning collection in a particular year.

The period between seed maturation and seed dispersal is often short, whereas the effects of climate in a given year may displace the dates of seeding by several weeks from the average. In the temperate zone an early spring and dry summer can cause very early seed ripening, while strong, dry winds cause rapid dispersal of the ripe seeds. Cool, wet weather, on the other hand may delay ripening and dispersal by weeks or months. In the dry tropics there are similar annual variations in the dates of the onset of the dry season and of the rains. It is therefore necessary in each year to check the correct timing of collection by examination of the crop itself.

The reconnaissance of the *size* of the seed crop, made 1 – 2 months before seed collection as described on p. 28 will also give some indications as to how the seeds are maturing. Conclusions from it should be twofold e.g.: - "Stands A, B and C: Very light crops, not worth collecting this year. Stands X, Y, Z: Good crops, seeds probably mature in 4 weeks' time". A final check on the ripeness of seed must, however, be done at the time of collection.

The humid tropics present special problems because seasonality effects are usually subtle or absent and the period of maximum seed production is uncertain. After detection of flowering in a stand from which it is desired to collect seeds, it is important to carry out periodic reconnaissance to check on the progress of fruit maturation. An efficient schedule of reconnaissance requires prior knowledge of the length of time between anthesis (flower opening/pollination) and fruit maturity.

Among Malaysian trees, the period from anthesis to fruit maturity varies from 3 weeks for *Pterocymbium javanicum* to 11 months for *Diospyros maingayi*. In the exotic Brazil nut, *Bertholletia excelsa*, the period is 15–16 months. In Malaysia it is recommended that if the maturation period is X weeks, the development of the crop should be checked at 1/2 X and 3/4 X weeks after flowering. A fixed and arbitrary schedule of, say, once a month, will result in the collector being too late for a fast-maturing fruit like *Pterocymbium javanicum* and too wasteful of effort in the case of a slow-maturing fruit like the Brazil nut.

Apart from the exceptional case of deliberate collection of immature seeds, seed collectors need to be able to time collection for the period when seeds (but not necessarily fruits) are fully ripe but before they are dispersed by fruit dehiscence or fruit consumption by animals. To achieve this aim collectors must be able to distinguish ripe from unripe seeds. Several different methods have been used for the recognition of seed maturity. None of them work perfectly on all species and a good deal of experience or research is needed to determine the best method, or combination of methods, for a hitherto unfamiliar species. They may be divided into those which are of direct application in the field and those which need laboratory equipment. The latter may be valuable in providing a check on the field methods, but are unlikely to be of practical use to the collector unless the collection site happens to be close to the laboratory, which may be the case for some seed orchards.

Laboratory Methods

a) *Dry weight*. The most generally accepted measure of maturity is the time when the seed has reached its maximum *dry* weight, a point called physiological maturity. This means that nutrients are no longer flowing into the seed from the mother tree. Maximum *fresh* weight does not indicate physiological maturity because the maturing seed begins losing water while nutrients are still being accumulated and biochemical processes are continuing.

   Recurrent dry-weight determination of a series of seed samples can be made and the results extrapolated to the remaining crop, but this method is slow and therefore seldom used.

b) *Chemical analysis*: Biochemical changes take place as seeds mature but relatively little is known for most species. Chemical indices of seed maturity have been determined in a few species e.g. content of crude fat and protein-nitrogen, which increase five and four times respectively from immaturity to physiological maturity, are the best chemical indices for *Fraxinus pennsylvanica*. But they have no advantages over an examination of the embryo and the colour change of the fruit, and the extra trouble to perfom the analyses does not appear to be justified. Rediske found that *Pseudotsuga* seeds were physiologically mature when the content of reducing sugars dropped to 14 mg/g.

c) *X-ray radiography*: The examination of the development of the embryo and endosperm of sample seeds by means of X-ray radiographs is a quick and relatively straightforward method of assessing seed maturity, provided that suitable facilities and skilled technical staff are available. The technique has been used successfully for *Tectona* and a number of other tropical species, as well as for temperate species such as *Pinus strobus*. It has the disadvantage of requiring relatively expensive equipment and relies heavily on the judgment of the seed analyst for reliable results.

d) *Moisture content of fruits*: Water loss of maturing cones and fruits occurs in many species and is closely related to the maturity of the seed. Seeds of *Picea glauca* are considered ripe when moisture content falls below 48 %, of *Larix decidua* at 25–30 % and of *Pinus sylvestris* when it falls to 43–45 % (fresh weight basis). However, determination of moisture content by drying in an oven suffers the same disadvantage of slowness as determination of dry weight.

Field methods

e) *Specific gravity of fruits*: As moisture content of fruits and cones decreases with maturation, so does specific gravity or density, the ratio of unit weight to unit volume, decrease. Unlike moisture content, it is not too difficult to determine approximate specific gravity in the field by flotation in liquids of known specific gravity. Specific gravity indices of maturity have been established for cones of a number of coniferous species, and the cone to be tested is placed in a liquid in which it will float if mature and sink if immature. Various mixtures of kerosene (SG = 0.80), light SAE 20 motor oil (SG = 0.88) and linseed oil (SG = 0.93) have been used to prepare flotation liquids having a designated specific gravity. Tests must be made immediately after cones are picked from a tree. Specific gravity indices have proved reliable for some temperate conifers e.g. an S.G. of 0.74 for *Picea glauca*, but not for several southern hardwoods in the USA.

f) *Examination of seed contents*: Examination of seed contents exposed by cutting open fruits or cones lengthwise can be a reliable and simple method of assessing seed ripeness, provided the operator is experienced. Most embryos and endosperm pass through an immature "milk" stage, followed by a "dough" stage when the tissue becomes more firm. Mature seeds have a firm white endosperm (where present) and a fully developed firm embryo.

g) *Colour of fruits or cones*: Colour changes in fruit or cone provide a simple and, in some species, reliable criterion for judging seed maturity, but the operator must be experienced in the characteristics of the species concerned. In common with the specific

> gravity method, it involves no destruction of the seeds in the sample examined. Colour changes are usually from the green of the immature fruit or cone to various shades of yellow, brown or grey, and this may be accompanied by hardening of cone scales or of the pericarp of dehiscent or woody fruits. Since the seed normally matures before the fruit, it is advisable in some species to time collection at an earlier rather than a later stage of the colour change. Colour change was found to be the most reliable indicator of maturity for general practice in several southern hardwoods in the USA. It has also given good results in a number of temperate conifers. In Malaysia Tamari found that the best results were obtained by timing collection of Dipterocarp fruits when the wings turned brown but before the fruit itself changed colour.

In Thailand cone colour is used as a guide to optimum time of collection of pines, but differs according to species. In *Pinus kesiya* collection starts when cones have hardened and the colour is changing from green to brown in proportions of 50 : 50. In *Pinus merkusii* optimum time of collection is reached when the majority of cones are brownish and some have started to open. Trials with the Zambales (Philippines) provenance of *P. merkusii* have shown not only that extraction is a much more lengthy and expensive operation with green than with brown cones but also that the seed extracted has lower germination rate. Experience with *P. caribaea* in Honduras is similar.

Abscission and shedding of fruits is usually a sign of fruit maturity and it might be assumed that it also indicates a high content of sound, mature seeds. This is not always the case. The first seeds or fruits which fall naturally are often of poor quality, in which case it is advisable to reject them and to postpone collection until the peak and the latter half of the season. In Thailand fruits of *Tectona grandis* start to be shed in March but observations have shown that the most viable fruits are the last to be shed, so collection starting only in April is recommended. The first fruits of Dipterocarp species that fall upon ripening are usually defective and collecting should be delayed until the greater portion of the fruit has fallen.

Collection of Immature Seeds

It is general practice to collect seeds when they are mature, because they have a higher germinative energy and a greater longevity in storage than immature seeds. An alternative method is to collect fruits prior to ripening and to store them in relatively cool, well ventilated conditions which permit afterripening of the seeds within the fruit. It has shown promise on a research scale in a number of species.

There are several reasons for the interest in developing techniques for artificial ripening. They are: -

— *to extend the collection season*: The short period between seed maturity and dispersal may place an excessive demand on the availability of seasonal labour and in some areas unfavourable weather conditions in the collection period may aggravate this situation. Lengthening the period available for collection permits better organisation of the collections and allows skilled personnel to pick more of the crop. It can be particularly valuable in research, where a large number of seed lots have to be collected from widely scattered localities. Griffin used the technique in his study of Douglas fir provenances.

— *to avoid damage to the seed crop by insects and other pests*: Insects, birds, rodents and other pests frequently damage or destroy seeds and fruits when they have reached maturity. Early collection may be one method of avoiding these losses. Damage and deterioration of seeds is usually worse during the period while they lie on the forest floor than during any other stage in their history; any reduction in this period will improve their subsequent viability and longevity in storage.

— *to salvage immature seed collected inadvertently*: Untrained collectors of seeds often begin picking fruits and cones too early in the year before they are fully mature. Artificial ripening provides a method for handling this material.

The development of techniques for after-ripening of immature seeds will require more research, before they can be applied to a wide range of species. However, where a problem of rapid dispersal or of seed pests exists and provided the earliest time for safe collection of immature fruits can be established, such techniques can be very beneficial.

Determining Seed Collection

If it can be assumed that the seed collector has received clear directions from the seed user as to which species and provenances and, in some cases, from which stands he is to collect, it is still his responsibility to select the individual trees for collection. Criteria will vary considerably according to whether the collections are large-scale for afforestation projects or small-scale for research purposes.

Identification of species presents no problems in monospecific plantations, but is essential and may be difficult in mixed natural forest, especially where very similar species of the same genus occur mixed together as happens with pines in Mexico and Central America, eucalypts in Australia and dipterocarps in S.E. Asia. Unless identification is certain, it is often advisable to collect herbarium specimens as well as seed.

Large-scale Collections

In large-scale collections emphasis is on collecting as much seed as possible, as quickly and cheaply as possible, rather than on very careful selection of the parent trees. It is, nevertheless, essential to avoid collecting seeds from very poor phenotypes or seeds which prove to be empty or non-viable. Useful guidelines are listed by Stein et al., and form the basis of the following: -

1. Collect seed only from healthy vigorous trees of reasonably good form that are making average or better growth.
2. Where possible, collect from mature or nearly mature trees. Overmature trees should be avoided, since seeds from them may be of low viability.
3. Avoid isolated trees of naturally cross-pollinating species, since these are likely to be self-pollinated. Seeds are likely to be few, of low viability, and any seedlings produced are frequently weak or malformed.
4. Avoid collecting in stands containing numerous poorly formed, excessively limby, off-colour, abnormal or diseased trees.

Often it will be necessary to compromise between seed production and phenotypic appearence. No seed should be collected from excessively coarse-branched, vigorous "wolf" trees, even though they often bear a large crop, while trees of exceptionally good form sometimes bear so little seed that they do not warrant the trouble of collection. The bulk of seed will come from trees which are "average or better than average" in both form and seed production. Although few studies have been made of the reproductive biology of tropical trees, the occurrence of some species at a very low stocking (less than one every km^2) suggests that these must be naturally self-pollinated. Seed collection from such trees is free from disadvantages associated with collections from isolated trees of natural cross-pollinators.

Small-scale Research Collections

In small-scale collections for research, selection of trees will depend on the precise objective of the planned research. Provenance research is now receiving a good deal of attention in many countries. Advice of IUFRO on provenance seed collections includes the following recommendations on collections from individual trees: -

1. Collect from not worse than dominant and co-dominant trees of average quality, within "normal" rather than "plus" stands. Collections from superior phenotypes, if made, should be kept separately.
2. Collect from a minimum of 10 trees, preferably from 25 to 50 in the stand. If the stand is very variable, increase the number of trees. Record the number of trees and the approximate percentage which they form of the stand.
3. Seed trees to be at least seed fall distance apart from each other. A distance of 100 m has been

adopted for *Pseudotsuga*. This is to reduce the risk of collecting from half-sib parents. In Australia a minimum distance of twice tree height is used as a practical rule of thumb.

4. Individual seed trees to be marked.
5. Collect equal numbers of cones, fruits or seeds per tree.
6. In normal first stage provenance collections, seed from individual trees may be mixed together. If special studies on individual genotypes are to be done, seed from each tree should be kept separate.

Single Tree Collections

Foresters are interested in the variation within populations and provenance as well as the variation between them. In the case of exotics, one approach for the introducing countries is to study between provenance differences under local conditions first, and to investigate variation between individuals in the best provenance, by means of progeny trials, only at a later stage after the provenances locally best adapted have been identified. If progeny trials are the object of seed collection, it is essential to keep seed from individual trees separate at all stages of collection, transport, processing, nursery and field planting.

The preservation of the identity of individual trees through the collection and extraction phase often requires considerably more effort than bulking the collections. If the effort is made, then certain advantages accrue, as listed by Turnbull:

— It permits biosystematic study of genetic variation both within and between populations. McElwee states that seed collections *must* be kept separate by individual trees throughout the test from collection to outplanting, as provenance tests are weakened by combining seed within a stand, making it impossible to distinguish between seed source and individual variation. However, in large provenance tests with many individual trees, it may be beyond the resources of the investigator to retain the identity of the parents.

— It is possible to manipulate to equalise the amount of viable seed from each tree in the provenance mix if the seed must be bulked prior to sowing.

— It is not always possible to detect trees bearing hybrid seeds in the field. This is particularly the case with eucalypts. If seedlots are kept separate then, following the raising of small samples from each tree, any showing evidence of hybridisation can be eliminated before the main trial is established.

Single Clone Collections

In seed collection in clonal seed orchards, the unit of identity to be kept separate is often the clone rather than the individual ramet. In Zimbabwe the practice of keeping individual clonal seedlots separate has been followed for many years and it is considered that it more than justifies the additional costs and efforts involved over bulked orchard collections. The advantages can be summarised as follows:

a) Maintaining individual clonal identities at all stages from collections through to storage enables one to act with the minimum of delay on most information as it becomes available. The more obvious areas where action can be taken quickly are when rogueing becomes necessary or when susceptibility to pests or diseases becomes apparent, and the undesirable seedlots can be isolated or discarded.

b) Seedlots can be made up and supplied to suit specific sites, making use of the most recent information on genotype/environment interaction from progeny test analyses.

c) In conjunction with (b) above, and using the test results of each clone, composite seedlots can be made up in such a way as to give equal clonal representation in the final planting stock, avoiding the dominating effects that some clones frequently have in a bulked seedlot.

d) In general, it increases the options available to the user.

Collections for Conservation

Collections are also made for the purpose of attempting to conserve the gene pool *ex situ* either as seed in long-term storage or in planted conservation stands. Because exact knowledge of gene frequencies in the indigenous populations is largely lacking, collecting for gene conservation must be based mainly on commonsense. Similar methods as for provenance collections are likely to be appropriate, with the exceptions that: -

1. A somewhat larger number of trees per genepool should be sampled. Estimates are in the range or 50 to 100.
2. The sample should be a strictly random one and include poorer than average as well as better than average trees, in order to capture as much as possible of the total genetic variation. The only

restriction on this principle is the impossibility of sampling trees which are bearing no seed.

3. An additional measure to ensure the fullest possible genetic diversity within the seed collected is to collect in a better than average seed year. The better the seed year, the better the representation of male parents contributing pollen as well as of female parents supplying seed.
4. The quantity of seed to be collected from each provenance will usually be larger, because the recommended area for a conservation stand, 10 ha, is much greater than the total area of one provenance in a single provenance trial.

Assembling Resources

One part of planning is the timely assembly of clear information on the nature and magnitude of the seed collection tasks - number of species and provenances, seed quantities, location of stands, best dates to collect etc., as described above. The other part is to select and assemble the resources needed to do the job. During the planning stage, the leader of collecting operations needs to check the preparations for field work as follows:

1. *Organisation of collecting teams:* Known or estimated output of collecting teams needs to be related to the quantity of seed, number of stands and length of season, in order to determine the required number and size of teams. For example provenance collections of *Pinus kesiya* totalling 3000 kg of cones from 16 stands can be completed in 30 days by one team in Thailand. If the full season is estimated at 45 days and there is a demand for a total at 9000 kg of cones from the same 16 stands, two teams will be

needed to do the job. If planning can be done sufficiently in advance, it may give an opportunity to train additional climbers if they are needed. It is desirable to have at least one tree-climber on the permanent staff, who can be responsible for looking after climbing equipment and for training new temporary climbers. In the field the climbers should be organised in small teams with a foreman in charge of each. In Honduras a team of 6 pairs (with one climber and one ground assistant or anchorman in each pair) has been found to be a suitable size.

2. *Organisation of transport*: Collecting teams need to cut to a minimum the time spent in moving between one site and the next. Transport must be available where and when it is needed. If necessary, extra vehicles may be temporarily hired. In roadless country, advance arrangements may be needed to employ extra unskilled workers to assist in carrying equipment, tents, etc.
3. *Organisation of equipment*: Choice of equipment will vary greatly according to local conditions. The steeper and less accessible the terrain, the simpler and lighter should be the equipment. Whereas highly mechanised equipment such as tree shakers or mechanised platforms may be appropriate in large seed orchards on flat land, only light portable equipment is practical where natural stands are 4 – 6 hours' walk from the nearest road. Apart from collecting tools, safety clothing, first aid equipment and plenty of bags and sacks should be provided.
4. *Organisation of records*: Meticulous recording and labelling is essential to good collecting. Appropriate labels and forms need to be designed well in advance and printed in adequate numbers.

5. *Organisation of permits*: These are not normally required for forest services collecting in government forest reserves, but may be needed when collecting on private land, in National Parks and special reserves, or in another country. Even if formal authority is not needed, it is often advisable to inform local communities of proposed operations in advance.
6. *Organisation of seed extraction*: Arrangements for rapid movement of fruits from collection site to extractory may be needed, involving the advance organisation of transport. The seed extractory staff must be advised when to expect the fruits. If some preliminary sun-drying of fruits in the forest is planned, polythene sheeting or tarpaulins will be needed.

4

Seed Handling, Collection and Processing

During the period immediately after collection from the tree, seeds are particularly susceptible to damage. At the same time the environment in which they are placed, which is fairly easy to control in the seed centre or seed processing depot, is difficult to control in the forest and during transport from forest to seed depot. Fluctuations of climate cannot be predicted or prevented and transport may involve persons who do not have the same personal interest in the welfare of the seeds as the collector, processor or user.

During this period there are serious dangers of loss of the *identity* as well as of the *viability* of the material. The risks are especially high in many tropical countries, where temperature and humidity are high and where transport may be difficult, slow and uncertain. Investigations into the problems of storing "difficult" species have sometimes revealed that the real trouble occurs between collection and processing/storing. If seeds have already lost some of their viability before storage, even the best storage treatment will give poor results. Careful advance planning is therefore essential in order to provide the closest possible control over the identity and health of the seed at all stages in its movement.

Viability Maintenance

Almost invariably it is fruits, not seeds, which are picked from trees. Sundrying of fruits and extraction of seeds is carried out in the field in some countries where the climate is suitable. In others it is considered preferable to transport the fruits as quickly as possible to the seed processing depot, where the conditions of extraction can be controlled much more closely than in the field.

If seeds are not extracted in the field, great care must be taken of the fruits both in the forest and during transport. Bulk quantities of fruits in high temperature and humidity are very susceptible to deterioration through the action of moulds and other fungi and through overheating due to a high rate of respiration. The importance of good ventilation in reducing these dangers cannot be overemphasized. If fruits are stored temporarily in separate containers, they should not be filled to the top. In particular, sacks containing fresh cones should be only half-filled; in this way space is left for expansion of scales as cones dry. Otherwise, scales may acquire a set which severely impairs subsequent seed extraction. To facilitate air circulation within sacks, as well as for easy handling during transport, it is advisable to put only 10 – 20 kg of fruits in each sack. Loose-weave hessian sacks or nylon-mesh laundry bags allow good air circulation through the side of the container.

In the case of eucalypt capsules and fruits of other species with very small seeds, however, close-weave cotton bags should be used if there is any chance of the fruits opening during transit. Large open-mesh baskets are ideal for promoting free air circulation in cones and other large fruits and may be constructed from locally available materials, whether metal, willow, bamboo or rattan. Aeration in loose-piled fruits can be improved by inserting loosely-constructed "chimneys" of wooden slats in the middle of the piles. Daily turning of loose-piled fruits or of sacks can do much to improve access of air to the less exposed fruits.

Table 1. Interim cone storage racks are common in North America

If fruits cannot be transported at once to the seed processing depot, temporary field storage must be arranged, in sheds or under some kind of shelter. Shelter is needed against rain and, for some species, against too high insolation. Sheds should be open-sided or otherwise well ventilated and sacks well spaced on racks or hung from hooks to allow free air circulation. Hanging from hooks has the added advantage of giving protection against rodents. If storage is in the open, overhead shelter may be provided by canvas tarpaulins or polyethylene sheeting. If the collecting season coincides with a period of reliably dry but not too hot weather, no overhead shelter is necessary. Sacks should never be piled on top of each other in large heaps.

For most orthodox seeds some degree of advance drying of the fruits in the field is a good thing. Drying of orthodox seeds to under 12 % before shipment by air has been recommended. Drying can be facilitated by the use of open-weave sacks. Polyethylene bags are not suitable for temporary storage of fruits of these species, since they prevent drying and may encourage fungal moulds and overheating. However, fruits of recalcitrant species must be kept cool and moist to maintain viability of their seeds. Polyethylene bags, which prevent drying, are suitable as containers for species of this type.

For large collections it has been contended by Isaacs that large wirebound pallet boxes, with a capacity of 7.3 hectolitres, are much more efficient for the physical handling of cones prior to processing. These boxes, used as a single container for handling, shipping and storing, are delivered as components which can be assembled in less than three minutes as needed. But, because of their weight of around half a ton when filled, they need the use of a forklift truck for loading and unloading and are therefore only suitable for large and highly mechanized operations.

Special measures may be needed to prevent damage from pests and diseases. The use of insecticidal or fungicidal dusts may be advisable in some circumstances, if there is a high risk of severe damage, but great care is needed in treating fresh and relatively moist seed, to avoid damage from the chemicals themselves. Maintaining fruit hygiene, particularly through good ventilation, is usually preferable to reliance on chemicals. Storage of sacks off the ground will itself give some protection against rodents. The incidence of pests and diseases is often worst on the forest floor and prompt collection of fallen fruits can do much to minimize subsequent losses.

Seed Extraction Close to the Collection Site

The decision whether to extract the seed from the fruit at a central seed processing depot or close to the site of, and soon after, collection must be made in the light of local conditions. As stated by Kemp, reduction in bulk and weight of the collected material greatly facilitates transport and for this reason the extraction of the seed from relatively bulky fruits may be desirable at an early stage, even if the more critical operations of dewinging, cleaning and final drying are better done at the central depot. In some cases it has also been found that early extraction of the seed is essential to maintain maximum viability.

Fleshy fruits such as the syncarp of *Chlorophora* ferment if stored in bulk for any length of time and the fermentation impairs seed viability. The use of the sun to speed extraction and drying of the seed is common in the tropics, particularly for many conifers and trees with woody, dehiscent fruits e.g. eucalypts, and is generally beneficial. However, too rapid drying of fruits may sometimes cause them to remain closed and prevent later extraction, a condition analogous to the "case-hardening" of timber. This is true of the cones of some pines if they are collected before peak maturity.

In such cases it is necessary to keep the fruits under cover, with a good circulation of air, for one or two weeks before more rapid drying is attempted. Some seeds, particularly of species from the ever-wet tropical forests, are easily killed by rapid drying of the fruits. For these species sun drying of the fruits is inappropriate and they should instead be kept moist during transit. Factors to be considered when deciding whether to extract locally or centrally include:

1. Distance to central seed processing depot and efficiency of the transport system. The longer the period to be spent in transit, the greater the risk of deterioration, the greater the saving in cost from transporting seeds instead of fruits, and the stronger the argument for local extraction.
2. The characteristics of the species in question. For example, in fleshy fruits with a high risk of fermentation, seed should be extracted by maceration locally and soon after collection while dry, resistant fruits and seeds of some leguminous species can withstand a lengthy transit period under poor conditions without serious loss of viability.
3. The suitability of the species for extraction by sun. Species which need kiln treatment for full extraction of the seeds will usually be sent to the central kiln-drying facilities without any attempt at local drying, but use of portable kilns or artificial heat from sawmills is sometimes feasible. Recalcitrant species which will not withstand drying are best despatched moist to the central seed depot as soon as possible.
4. The reliability of sunshine during the collection season. Dry season collection in Mediterranean or dry tropical climates should ensure ideal conditions. Thus in Greece the present practice of field drying and extraction has proved successful for most species. In the ever-wet tropics, in cool-

temperate conditions and for collections during the rainy season, sundrying is not practicable.

Even where seed extraction is done locally, it is preferable to do it where there are some facilities in buildings and communications rather than in the forest itself. In a country with a well-developed forest service, the local Forest District HQ is often the best place. Seed collecting in inaccessible and sparsely inhabited forests will have no such facilities and will need to improvise a seed extracting depot in the forest.

Identity Maintenance

To ensure maintenance of seed lot identity, each container of fruit must be labelled correctly when it is filled. As an additional insurance against accidental loss of the exterior label, identical labels should be placed both inside and outside the container. Weatherproof labels should be used and the minimum information recorded should include species, seed lot number, geographic location or name of seed source and weight of seed contained, date of collection and collector's name. Information on seed lot number and species is the key for seed documentation. If detailed information is given on seed collection data sheets the information given on the labels can be limited to seed lot number, species, seed source and weight of seed contained. A copy of the seed collection data sheet or certificate of origin with reference to seed lot number must be attached to the documents accompanying the seed - or preferably be mailed before the seed is despatched. When one seed lot is divided between several containers for despatch, each label should state also the number of containers involved .

In large-scale operations, information may be coded. In the case of small collections for research purposes, e.g.

provenance collections or single-tree collections for progeny trials, additional information is recorded on a separate certificate of seed origin or collection data sheet. Reference to seed lot number is important. Labelling is more time-consuming, but also it is even more important, in numerous small research seed lots than in bulk collection, since it is essential to keep each seed lot separate and clearly identified at all stages between collection and sowing in the nursery. Small nylon bags are ideal containers for small lots of cones and seeds which need to be kept separate. They allow good ventilation and many of the stages of seed handling (transport, sun or kiln drying manual dewinging) can be carried out without removal of the cones or seeds from the bag.

Additional information, apart from that on the label, often needs to be recorded, especially for research seed collections. There are three main purposes for documentation of seed, (i) to record the location of the collection, so that good sources can be revisited if necessary at any time in the future and, equally important, bad sources avoided, (ii) to provide information on the ecological conditions, the actual populations sampled and the methods of collection and handling of the seed, to assist with interpretation of research results or the planning and conduct of other collections and (iii) to comply with requirements for safe and rapid transport and acceptance of the seed. Examples of seed collection recording forms are included in Appendix 1. Special documentation which may need to accompany seeds in transit, especially if movement is between countries, includes collecting licences, seed movement orders, export and import permits, phytosanitary certificates and certificates of origin or of genetic value under national or international schemes of seed certification.

Transport

It is important to ensure that the minimum possible time should elapse between the despatch of fruits or extracted seed from the collecting site or local seed collecting depot and its arrival at the central seed processing depot. Part of the planning operation must be to provide transport which is adequate in quantity and quality to avoid delays in despatch and breakdowns en route. Even short stop-overs add to the heat build-up in fruit and cone sacks during transit. Vehicle drivers should be informed of the nature of their loads and the need for proper care and prompt delivery. Road transport is commonly used for at least the first part of the journey. Railway transport may be more economical over long distances and air transport quicker, but both involve some loss of control over storage conditions in transit and any trans-shipment means extra handling and delay. Road transport is likely to be the preferred method in most conditions.

In large-scale seed collection operations in *Gmelina arborea* plantations in the Jari project in Brazil, great emphasis is placed on reducing the period in transit in sacks to the minimum. The average period is 1.2 days. Germination of fruits may drop by 22 percentage points after a day and nearly to zero after a week. In some areas seed maturity and weather conditions affect the safe period in transit. In Zimbabwe pine cones collected early in the season, at the height of the rains and when seed moisture content is high, should not remain in sacks longer than 2 – 3 days; drier cones collected towards the end of the rains may remain up to 10 days without deterioration.

If transport distances are short and large quantities of a single species and provenance are to be transported, fruits may be loaded into vehicles without containers. The vehicle must be cleaned of all seed carried on previous trips before a new lot is loaded. For longer journeys or smaller seed lots individual containers should be used. Sacks should be carefully arranged in the vehicle to allow as much air circulation as possible between them. Open-mesh baskets are excellent for promoting free air circulation during transport as well as during temporary storage.

For most species open trucks and trailers should be used in preference to closed vans in order to promote air circulation. But for species which need to maintain a high moisture content if they are to retain viability, care must be taken to prevent excessive drying; use of polythene bags and provision of shade against insolation are necessary. Rapid transport direct to the final destination and immediately after collection may be essential for some species which either germinate naturally or lose their viability soon after seed fall at normal temperatures.

Provision of special insulated containers, to control temperature and humidity in transit, may also be needed. Styrofoam containers or metal thermos flasks are recommended as a precaution against temporary exposure of recalcitrant seeds to cold or freezing conditions in aircraft holds. This situation may be found among some trees of the ever-wet tropical high forest but is rare among species which have evolved under conditions which include a severe cold season or dry season each year. Many of the species most widely used in plantation forestry fall into the latter category.

With valuable collections it may be advisable to split each lot of seed or fruits into at least two parts, to travel separately, in order to guard against losing an entire collection through accidents en route. It may also be desirable to insure the contents against loss or damage, for a sum that will cover at least a proportion of the cost of repeating the collection. Information on the estimated time of arrival of a shipment of fruits must be sent in advance to the seed processing depot. This enables the receiving station to provide staff for prompt unloading of fruits. Similar information should be sent to intermediate recipients responsible for transshipping or forwarding consignments of fruits.

Precautions for Recalcitrant Seeds

Most of the problems in maintaining seed viability which are outlined in this chapter are accentuated in the case of recalcitrant species in the humid tropics. They have short life spans, and can neither tolerate low temperatures (not much below 20°C) nor reduction in their moisture content below a relatively high value. The majority of seeds in the humid tropics are recalcitrant, and as a result of high rates of seed spoilage in transit, such species are seldom used in reforestation except within their countries of origin. Even for seeds collected and used locally, deterioration can be serious within a matter of days unless special precautions are taken. Collectors have to work within relatively narrow limits of tolerance. The main precautions to note are as follows:

— *Ventilation*: Recalcitrant seeds (and their fruits) respire actively, hence require good ventilation. If large quantities are closely packed, suffocation, physiological breakdown, fungal growth, and overheating will occur resulting in rapid death of the seeds. If plastic bags are used as containers, their tops should be left open or small holes should be made in their sides. Baskets or cloth bags are suitable although usually more bulky or expensive. It is not easy to strike the right balance between adequate ventilation and adequate moisture conservation.

— *Temperature*: Temperatures below 20°C or above 35°C should be avoided. Low temperatures are likely to be experienced in air transportation unless the seeds are kept in the pressurized cabin. High temperatures may be due to respiration or to direct heating by the sun. Good ventilation will reduce heat build-up from respiration. Recalcitrant seeds should be kept shaded from direct sun at all times.

— *Moisture content*: Recalcitrant seeds deteriorate if their moisture content is reduced too much or too rapidly. This is likely to happen during transportation in open vehicles because of air movement. The size and number of ventilation holes in the containers should be reduced under such circumstances. Open containers should be covered with newsprint or cloth to reduce the dessicating effect of air movement.

— *Organization of nursery*: Before the collection is made, the recipient nurseries should be forewarned to have their germination beds ready. Recalcitrant seeds should be sown as soon as possible after collection.

— *Long trips*: Collection trips for recalcitrant seed should not exceed a few days in duration. If a long trip cannot be avoided, a lot of extra work has to be done in daily inspection and processing of the collections already made. If decay and fungal growth set in, the seeds will have to be spread out for better ventilation. Decaying pulpy fruits will have to be separated from the healthy fruits and depulped immediately. Capsules must be discarded as soon as they have opened sufficiently for their seeds to be extracted. If seeds begin to germinate during the trip, the germinants may be saved by storing in rigid containers or baskets lined with newsprint or other absorbent material and kept moist. Some seeds deteriorate so rapidly that the best way to transport them may be in a germinating condition in a moist medium.

5

Wheat Seed Production

Wheat is a high-volume, low-profit seed crop and has been produced primarily by heavily subsidised government seed programmes. With privatisation and liberalisation, many of these programmes are at risk of being closed down. The private sector, however, may not focus on wheat seed due to its characteristics (self-pollinating, high-volume and low-profit). If private seed enterprises exist, they consider wheat seed to be of secondary importance. Furthermore, in most countries there has been no ongoing effort to promote the use of improved seed by wheat farmers, and no significant breeding developments have recently taken place to increase yield and quality.

Since wheat is a self-pollinating crop and the grain can be used as seed, farmers tend to replant their own seed. It is, therefore, expected that in the future the large majority of resource-poor, small-scale farmers in many developing countries will have to rely on seed saved from the previous harvest.

In the developed world, declining world market prices for grain do not encourage farmers to produce wheat. Moreover, there have been no widespread disease outbreaks, which make seed treatment (and thus purchase of certified seed) necessary. Countries in Europe and the United States are attempting to reduce subsidies, which will result in further unwillingness to invest.

On the other hand, there are possibilities in developing countries for major increases if significantly improved varieties appear on the market, disease outbreaks occur and organised promotional efforts emphasise maximum production efficiency including improved seed. Introduction of Plant Variety Protection and/or sophisticated technologies, such as Genetic Use Restriction Technology (GURT), may also induce the private sector to invest in wheat seed.

In Australia, Europe and the United States, there is a possibility of increased use of seed of higher yielding varieties if subsidy reductions and/or market trends cause a decline in production area, with only the more efficient farmers remaining in production and attempting to produce more (or the same total amount) and more efficiently from less area as a result of trying to diversify crop production.

In modern agriculture, seed is a vehicle to deliver almost all agriculture-based technological innovations to farmers so that they can exploit the genetic potential of new varieties. The availability, access and use of seed of adaptable modern varieties is, therefore, determinant to the efficiency and productivity of other packages (irrigation, fertilisers, pesticides) in increasing crop production to enhance food security and alleviating rural poverty in developing countries.

For seed to play a catalytic role, it should reach farmers in a good quality state, i.e. high genetic purity and identity, as well as high physical, physiological and health quality. In contrast to fertilisers and pesticides, farmers select and save seed to plant the next year's crop, and any off-farm seed from the formal sector should be of a better quality for farmers to invest in it. Therefore, the best production techniques need to be followed to produce good quality seed.

For wheat, seed and grain production follow rather similar operations but different strategies. Apart from good agronomic management of the crop, seed production differs from grain production on the following key issues: land requirement, isolation, roguing, prevention of contamination and limitations of generations. Another difference is that seed crops must meet specific quality standards prescribed by the national seed regulations. The technical, administrative and legislative control by the certification agency provides guidelines that have to be followed to produce good quality seed that meets the standards.

Evaluation of New Variety Development

Modern varieties are the backbone of the formal seed industry. National Agricultural Research Systems (NARSs) have a major responsibility for variety development and for generating appropriate technologies to better utilise the yield potential of new varieties. The new varieties must pass through a series of evaluation, release and registration tests and procedures before farmers can use them. Basically two type of tests are carried out: Distinctness, Uniformity and Stability (DUS) and Value for Cultivation and Use (VCU). A DUS test is a descriptive assessment that establishes the identity of the new variety, by using morphological characters, as well as its uniformity and stability. It is a useful tool for the purposes of seed production, certification and plant variety protection. The DUS tests usually run for two years. The new variety is compared with existing varieties to establish its distinctness. A variety description is prepared, and differences with other varieties noted.

The VCU trials focus on the benefit of the variety to the end users, farmers and consumers. National multilocation variety trials enable identification of superior varieties that meet diverse agronomic and consumer requirements. The VCU tests usually run for three years. In some countries, the variety is tested in on-farm verification trials under farmers' management conditions during the last year. Based on DUS and VCU test results, a variety may be released and registered for farmers' use. Many developing countries give priority to agronomic (VCU) trials rather than descriptive (DUS) tests. While both tests are important, the benefits of the two tests must be considered based on the immediate need of the country to use available resources efficiently and economically.

Seed Classes

Seed production follows a generation system to ensure that all seed that is marketed to farmers originates from a known source (breeder seed). When a variety is officially released, the small amount of breeder seed received from the breeder (agricultural research centre) is multiplied through a number of generations before it becomes available to the farmers in larger quantities as certified seed. Each generation is produced under strict supervision and must meet seed quality standards. The number of generations that are allowed after breeder seed depends on the mode of reproduction of the crop, risk of contamination, multiplication ratio and quantity of the seed required. For wheat, four to five generations are commonly used.

Different generation schemes exist (Table 1), which vary very little, particularly in nomenclature. The procedures followed are essentially the same. The Organisation for Economic Cooperation and Development (OECD) generation scheme is used as outlined below:

- — Breeder seed is the initial source of seed and is usually produced by the breeder. It is the source for the production of pre-basic or basic seed.
- — Pre-basic seed is the progeny of the breeder seed and is usually produced under the supervision of a breeder or his designated agency. This generation is commonly used for crops that have low multiplication ratios and where large quantities of certified seed are required.
- — Basic seed is the progeny of breeder or pre-basic seed and is usually produced under the supervision of a breeder or his designated agency and under the control of a seed quality control agency.
- — Certified seed is the progeny of basic seed and is produced on contract with selected seed growers under the supervision of the seed enterprise, public or private. Certified seed can be used to produce further generations of certified seed or can be planted by farmers for grain production.

Breeder seed production is not monitored by the seed certification agency, while basic seed and certified seed are covered in the seed certification scheme. The seed quality control agency verifies the quality both in the field and in the laboratory and certifies that the seed meets the national standards. Such classes of seed are known as certified. It is important to note that all certified seed classes relate to a breeder seed through one or more generations.

Table 1.
Comparative seed nomenclature in selected countries of West Asia and North Africa

Definition	*OECD*	*AOSCA*	*Ethiopia*	*Egypt*	*Jordan*	*Algeria*
1st generation supplied by plant breeders	Breeder	Breeder	Breeder	Breeder	G_1	G_0 (G_1, G_2)
2nd generation	Pre-basic	Foundation	Pre-basic	Foundation	G_2	G_3
3rd generation	Basic	Registered	Basic	Registered	G_3	G_4
4th generation	Certified 1	Certified	Certified 1	Certified	Certified 1	Registered 1
5th generation	Certified 2	-	Certified 2	-	Certified 2	Registered 2

Some developing countries, where natural disasters such as drought are a common phenomenon, recognise a 'commercial seed' class to meet seed shortages in emergency situations. In such cases, the standards for certified seed are often lowered and accepted for distribution to farmers to overcome seed shortages. In other situations, commercial seed is simply a grain used as seed after laboratory testing for some quality attributes, such as purity and germination.

Variety Maintenance and Breeder Seed Production

Upon release of a new variety, a breeder will make available a small quantity of seed stock that is very pure and represents the variety. This stock is referred to as parental material and forms the basis of any future maintenance and seed multiplication of the variety. Laverack defines maintenance as "the perpetuation of a small stock of parental material through repeated multiplication following a precise procedure". For wheat, an ear-to-row method is recommended, where a number of ears (depending on the total quantity of certified seed required) that are true-to-type are selected, threshed separately and then planted in individual rows. During the entire growth period, the rows are inspected regularly, and any row with off-types or deviants is discarded. Ears are selected from the remaining rows to repeat the cycle, which is usually referred to as maintenance. The remaining rows are bulk harvested, and the seed is called breeder seed. In India, for example, 1 000 plants are used for wheat variety maintenance.

Each year the cycle is repeated to provide a regular supply of breeder seed for further multiplication to basic seed and then to certified seed. Maintenance and breeder seed production is the responsibility of the breeder or the institution that developed the variety. In many developing countries, maintenance is seldom carried out properly, and the responsibility is often taken over by seed production organisations.

Some national seed programmes (e.g. Ethiopia) have established special farms to produce early generations (pre-basic and basic seed) to maintain quality and availability. Laverack described different arrangements and management approaches for breeder seed production that can be useful and adopted in developing countries. In Morocco, a separate Seed Unit has been established within the Institut National de la Recherche Agronomique to maintain and produce breeder seed of public varieties.

Production Planning

New improved varieties developed by NARSs should be multiplied and made available to farmers in the shortest possible time to realise the benefits of investments in agricultural research. Appropriate seed production techniques coupled with strict quality control measures ensure that varietal purity and identity is maintained, which is the cornerstone of the entire seed programme. The rate at which the variety is multiplied and accessed restricts the availability of seed and its adoption and rapid diffusion through formal or informal channels.

Most farmers have a tendency to save their own seed of self-pollinating crops such as wheat compared to open-pollinated crops or hybrids where the risks of contamination and decrease in yield are higher. The seed renewal rate, the frequency in which farmers purchase certified seed of the same variety from the formal sector, varies between the type of seeds, crops, production environments and socio-economic factors.

The multiplication factor can be defined as the amount of seed harvested from each quantity of seed sown. It differs from crop to crop, but is largely dependent on climate, physical factors (soil fertility, etc.) and agronomic management. New varieties, after they enter commercial production, may lose their genetic potential or become susceptible to pests over time, which requires their replacement. Moreover, the varieties may also be exposed to genetic, mechanical and pathological contamination during the seed multiplication process. There is a practical need to limit the number of generations that the seed is multiplied after breeder seed.

For each generation, it is important to make an accurate estimate of the seed that needs to be produced every year. The total requirement depends on: (i) crop area planted; (ii) seed renewal rate; (iii) seed rate; (iv) multiplication ratio; and (v) number of generations. When planning seed production, the rates of rejection during field inspection, processing and laboratory testing should also be considered. Furthermore, a certain amount of carry-over seed should be taken into account, particularly for early generation seed.

Precise estimates of seed demand are difficult because several factors, such as weather conditions and socio-economic factors, influence farmers' decisions on which crop to grow the next year. In case of wheat, farmers may prefer to keep own-saved seed for a longer period than the anticipated seed renewal rates of four to five years. The seed demand of new varieties is even more difficult to estimate and depends on the adoption rate, which in turn is influenced by several technical, institutional, economical and sociological factors.

Production Arrangements

Generally, early generations (breeder and pre-basic seed)

are produced on agricultural research farms and basic seed on specialised seed farms, while certified seed is more conveniently and economically produced by farmers who are contracted by the seed enterprise (public or private). Farmers' cooperatives and associations are often involved in certified seed production. The seed enterprise provides technical guidance and advice to growers on various aspects of seed production, such as planting, roguing and harvesting, as well as monitoring and supervising field operations. The seed quality control agency inspects the fields and tests the seed in the laboratory.

In wheat, seed and grain crop production follow rather similar operations. The end product can either be used for consumption or for planting. Such similarity can easily lead to mixing seed and grain during planting, harvesting, transportation and storage. Seed production should be strictly monitored, and seed producers should be quality conscious. In developing countries, farmers are rigorously selected before they become seed growers.

In many developing countries, the availability of experienced seed growers is limited, and seed organisations provide considerable support. A village seed production scheme, which groups adjacent fields for seed production, will help to overcome problems of isolation distances and supervision when contracting smallholder farmers for seed production. The cost of seed production is higher than the cost of producing grain for consumption because it involves extra operations to maintain quality. Thus seed growers incur additional costs in field operations and management and must be paid a premium. In case of wheat, the additional premium paid over the grain price ranges from 10 to 20 percent.

Land selection

Wheat can be successfully grown in most parts of the world, both in tropical and temperate environments and on all soil types that are well drained and productive. Seed (certainly breeder seed) should be produced in areas: (i) where the variety is adapted; (ii) where soil conditions are optimal (to achieve a high multiplication ratio); and (iii) where climatic conditions are reliable to avoid loss due to natural hazards (flooding, drought, frost, etc.). The early generations should possibly be planted at two different locations to reduce the risk of losing the complete generation.

In some tropical countries, there is a high risk of rain coinciding with harvest time. Rainfall at maturity delays harvesting, causes sprouting and predisposes the crop to fungal attack, resulting in poor seed quality, i.e. reduced viability and vigour. Selection of appropriate sites is very essential to produce quality seed.

Previous cropping

The crop should be planted on a field with a known history to avoid contamination from volunteer plants, noxious weeds and soil-borne diseases that are potentially seed transmitted. The minimum number of years that is allowed between the planting of two seed crops (or seed and grain crop) is usually prescribed by the national seed regulations. Examples of requirements for previous cropping in countries in West Asia and North Africa are provided in the WANA Secretariat.

A wheat seed crop should never immediately follow wheat, unless the wheat crop in the previous season was of the same variety and of the same or higher generation. For basic seed, it is often recommended that the field should not have been planted with wheat or other small grain crops for at least two consecutive years. For certified seed, no wheat crop should have been grown in the previous year.

A suitable crop rotation plays an important role in pure seed production. For example, a two-year rotation for flag smut and seed gall nematode is suggested where applicable. For wheat, previous cropping could be legumes, vegetables or clean fallow, but other cereals (barley, oat, rye and triticale) and forage crops (oat) should be avoided.

Weeds

Seed contaminated with weeds could be the means for introduction and dissemination of noxious weeds. Mohamed mentioned that the introduction of wild oats in wheat in Egypt is attributed to contaminated seeds. Therefore, heavily infested fields with noxious weeds should be avoided. Unfortunately, fields in many countries are infested with wild oats (*Avena fatua, A. sterilis* and *A. ludoviciana*), noxious weeds that are spread almost all over the world and are difficult to eradicate. *Avena fatua* has no resistance to frost, and it is only problematic in summer wheat. In West Asia and North Africa (WANA), the problem is mainly *A. sterilis*.

Worldwide, a 10 percent potential yield loss in cereals is due to weeds, even with currently used control measures. In wheat, Tanner and Sahle cited that the use of clean seed had a significant effect on grain yield compared to weeding by hand or hoe. For land selection, the most essential pre-requisite is the selection of clean fields that are properly rotated and known to be free from contaminants.

Planting

Seedbed Preparation

Seedbed preparation is the same as for a grain crop. Wheat does not suppress weeds sufficiently and needs a clean, weed-free seed-bed for planting. In Sudan, weed competition studies on wheat showed that if weeding is not carried out between two and four weeks after sowing the yield is reduced by 20 percent.

Planting method

Planting with an automatic drill is recommended but not essential. However, row-planting has an advantage over broad-casting, as it requires less seed and facilitates mechanised weed control, roguing and field inspection. Roguing lanes (empty rows at intervals) should be left, which could be used by the seed grower to walk through the field when roguing and inspecting the crop, as well as for spraying the crop. Deep sowing delays emergence, resulting in weaker seedlings, reduced emergence and poor tillering and yield. Varieties with short coleoptile length, particularly semidwarf varieties, suffer most compared to varieties with longer coleoptile length.

Seed Rates

The optimum seed rates for wheat vary with variety, location and method of planting. For seed production fields, a lower seed rate may be recommended because lower seed rates lead to higher multiplication factors but to lower yield per unit area (Table 2). Higher multiplication factors lead to rapid seed increase, and farmers will benefit from the improved variety earlier. Low seed rates do not only increase the multiplication factor, but also often improve seed quality because a lower number of plants per unit of land receive better nutrition, thus producing better quality seed.

Table 2. Effect of seed rate on multiplication factor of wheat

Seed rate (kg/ha)	*Yield (kg/ha)*	*Multiplication ratio*
25	4 081	157
50	4 907	98
75	5 176	69
100	4 949	49
125	5 574	44

In practice, very low seed rates are not used. Lower seed rates may be used when planting early generations, but certified seed is planted at the normal or slightly lower seed rate. For certified seed, acreage becomes too large to be closely monitored and thus the risk of reduced yield is too high.

Seed Size

In wheat, seed size is positively correlated with seed vigour: larger seeds tend to produce more vigorous seedlings. Larger seeds of spring wheat produced higher yields than smaller seeds under late-sown conditions, but not under optimum management conditions. Similarly, Khah *et al.* found that low-vigour spring wheat seed produced lower yields only when it resulted in low plant populations or when planting was later than normal. However, Mian and Nafziger have found that seed size has little effect on emergence of soft red winter wheat.

Fertiliser Application

Fertiliser application to seed crops should be based on local recommendations. A well-balanced supply of nitrogen, phosphorus and potassium (NPK) is essential for seed production as it has an influence on seed development and thereby on seed quality. Phosphorus is essential for enhancing seed maturity and K for seed development. Ascher *et al.* concluded that seed nutrition combined with soil nutrition gave better yields and better seed quality. For example, cereals (wheat, barley and oats) grown from seed with a high P concentration gave 20 percent more yield than with low P, with a critical concentration of 0.30 percent for wheat seed. Ngugi reported that the use of compound fertilisers with sulphur (S) and calcium (Ca) gave higher grain yield and better seed quality (higher bushel weights) in non-traditional wheat-growing areas of Kenya.

The benefit of fertiliser is not always apparent. Fertiliser may increase the disease incidence and competition of weeds, such as wild oats. High N levels may promote vegetative growth, delayed maturity, predispose the crop to foliar diseases and lead to severe lodging and reduced yield and seed quality. In Ethiopia, both N and P increased the incidence of stripe rust, though the effect of P was less pronounced. In a similar study, the density of wild oat panicles increased with N fertiliser, but decreased with P fertiliser. Top dressing wheat crops with high levels of N will make the crop lodge, which makes it difficult to be field inspected by the seed certification service.

Cleanliness of machinery

Cleanliness of machinery upon planting is very important. Seed drills should be cleaned with compressed air when changing between varieties and other crops of similar seed characteristics. Vans and trailers used for transporting the seed should be completely clean to avoid contamination.

Isolation

Isolation, growing a seed crop separate from all sources

of contamination (genetic, physical and pathological), is one of the fundamental seed production techniques. In practice, such contamination can be reduced by not planting a seed crop in the vicinity of a similar crop that may contaminate it. Small, long and narrow fields need larger isolation distances than large or wide fields. Long and narrow strips are more prone to contamination than square fields. The minimum distance required for a particular crop is usually prescribed by the national seed regulations and depends on the seed class. Minimum isolation distances are larger for the early than for the later generations.

Table 3.
Examples of isolation distances for wheat in West Asia and North Africa

Country	*Isolation distance (m)*		
	Seed class		
	Pre-basic	*Basic*	*Certified*
Cyprus	-	2	2
Egypt	5	5	5
Iran	5	5	5
Sudan	4	4	4
Syria	2	2	2
Tunisia	1	1	1

Wheat is entirely a self-pollinating crop with a very low percentage of crosspollination, from 1 to 4 percent. Therefore, the risk from genetic contamination through out-crossing is small. Appropriate isolation is, however, required to minimise physical contamination. For wheat, it is usually sufficient to have a small strip of land between different fields to avoid mechanical admixtures. Minimum isolation distances (in metres) used in some Middle Eastern and North African countries are presented in Table 3. In India, an isolation distance of at least 3 m is required to separate a wheat seed field from the same variety not conforming to the standard.

Early generations, such as breeder seed, can be located in the middle of a field with the same variety, which is a common practice in some North African countries. Breeder seed (G_1) grown in the middle of a pre-basic (G_2) field and G_2 in the middle of a basic (G_3) field of the same variety to reduce risk of genetic and mechanical contamination (for names of generations refer to Table 1).

For wheat, isolation distances are more important when dealing with smut-susceptible varieties. In India, Agarwal reported that basic and certified seed fields of susceptible wheat varieties should be isolated by 150 m from fields infected with loose smut in excess of 0.1 and 0.5 percent, respectively. In Morocco, a 150 m isolation distance is required from other wheat fields if infection with loose smut is beyond 0.1 percent for pre-basic, 0.2 percent for basic, 0.3 percent for certified 1st generation and 0.5 percent for certified 2nd generation seed.

Restricting the number of varieties grown for seed multiplication per farm will reduce the chances of contamination. For example, in Tunisia only one wheat variety can be multiplied on a farm.

Crop Management

Agronomic management should be optimal and is similar to that for a grain crop. Small differences do, however, exist:

— using lower seed rates to increase the multiplication factor;

— leaving lanes to facilitate roguing and inspection;
— applying slightly less than optimum amount of N to reduce lodging;
— maintaining the species and variety purity;
— controlling diseases that are seed transmitted.

Roguing

Roguing, removing undesirable plants, is another fundamental aspect of seed production. Undesirable plants, commonly known as rogues, are: (i) off-types or genetic variants of the same variety; (ii) other varieties of the same species; (iii) other crop species of similar growth habit and seed characteristics; (iv) noxious weeds; and (v) infected plants with seed-borne diseases. Roguing is carried out to maintain the variety and species purity of the crop and to keep the seed crop free from seed-transmitted diseases.

Roguing follows a systematic procedure (Figure 1) and should be carried out at the time when rogues can be most easily identified and before any contamination occurs to the seed crop. The best periods for roguing a wheat seed crop are at heading and at maturity because off-types and other varieties of the same species are most easily identified. In wheat, all off-types, including those that appear due to residual segregation, should be removed.

For example, the inherent problem of tall off-types in semidwarf wheat varieties due to aneuploidy, arising from chromosome pairing failures at meiosis, has been discussed by Storlie and Talbert and Worland and Law. Similarly, other crop species whose seed is difficult to separate during processing (barley, oats, triticale, rye), noxious weeds (wild oats, lolium) and infected plants with seed-borne diseases (i.e. *Fusarium, Helminthosporium,*

Tilletia) are usually easily distinguished at these stages. It may be necessary to rogue the field several times, and all tillers and roots should be removed (pulled) to prevent regrowth. Rogues should be removed from the field and disposed of far away from the field or burned.

Roguing loose smut (*Ustilago tritici*) infected plants has no effect (except in reducing the disease inoculum) because the spores have already been spread and infection of the seed crop has taken place by the time the symptoms become visible. Therefore, certification schemes often do not allow the removal of smut-infected plants. However, reports from India indicated that loose smut-infected plants do head earlier and that flag leaves of smutted tillers exhibit yellow chlorotic streaks, yellow patches or yellowing at tips, even before ear emergence, and can thus be rogued before flowering. However, plants without streaks were also smutted.

Seed certification schemes set minimum standards for each class of contaminants that are permitted in a seed crop. Roguing aims at ensuring that these standards are met. Roguing should be carried out only when the seed fields do not meet these standards; roguing fields that meet the standards is not economical. While roguing, no selection should be carried out to ensure that the genetic make-up of the variety remains the same. Therefore, roguing of early generations should preferably be carried out by the breeder or under the breeder's supervision. Roguing in early generations can be most effective.

Mechanical Harvesting

Mechanical harvesting is a common practice for seed production fields. Breeder and pre-basic seed are harvested by plot combine and do not constitute many problems. Basic and certified seed, however, have to be

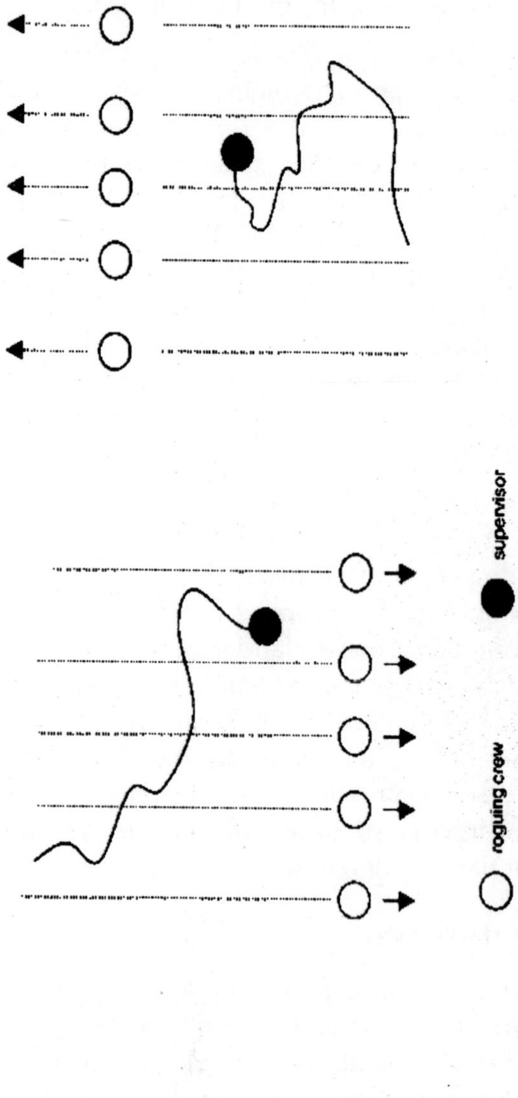

Figure 1. Diagram of haw a five-member roguing crew follows roguing through a seed field, followed by the supervisor

harvested with commercial combine harvesters. The most critical factors to be considered are seed moisture content, mechanical damage and cleanliness of equipment. For seed crops, dry weather during ripening and harvesting is essential.

Moisture

Cereal seed reaches physiological maturity between 35 to 45 percent moisture content, but it needs to dry down to a safer moisture content for harvesting and storage. The seed moisture content can be used as an indicator of when the crop is ready for harvest. Electric moisture meters or the crop characteristics can be used to decide when to harvest. For wheat, threshing or combine harvesting at 16 to 19 percent moisture content reduces mechanical damage.

Mechanical Damage

Proper adjustment of the concave clearance and drum speed of a combine is essential to avoid damage to the seed crop. Mechanical damage becomes a serious problem for durum wheat seed production in Algeria and Morocco. In Algeria, seed harvested by a single ear thresher or combine showed low germination and increased fungal attack.

Cleanliness

Combine harvesters are often difficult to clean and may still harbour contaminating seed even after thorough cleaning. The availability of compressed air is important. The combine should be thoroughly cleaned before harvesting, as well as between different varieties. When harvesting the next variety, the first few hundred kilograms of seed may be discarded because contamination may still be present in the combine. For larger plots of certified seed, it is also possible to harvest and discard the outlying rows of the field. After harvesting, the seed should be packaged in new and clean bags to avoid contamination.

Processing

After a seed crop has been harvested, the seed, if necessary, has to be dried and cleaned, i.e. removal of inert matter, seed of weeds, other crops and other varieties, and seeds that are diseased, damaged and deteriorated. Cleaning can be done because wheat seeds differ in length, width, thickness, density, weight and shape.

For wheat seed cleaning, mainly screens, indented cylinders and air are used. Screens separate based on the width and thickness; a width (or diameter) separation is obtained by round screens, while for thickness separation oblong screens are used. Indented cylinders carry out length separation; the indents (cells or pockets) in the cylinder will, depending on their size, lift the seeds, which fit in the indents.

Air separates seeds according to their behaviour in an air stream. The most important characteristic is the weight; light particles (dust, chaff, glumes or empty or partly filled seeds) will be lifted, whereas the heavier seed will fall down through the air stream. Each crop requires a different set of machines.

The raw (unprocessed) wheat seed is received from growers in bags or bulk and sampled to evaluate the need for fumigation, drying and cleaning, as well as to guide and monitor processing operations. If seed moisture is too high, the seed is first (pre-cleaned and) dried. If insects are present, the seed should be fumigated.

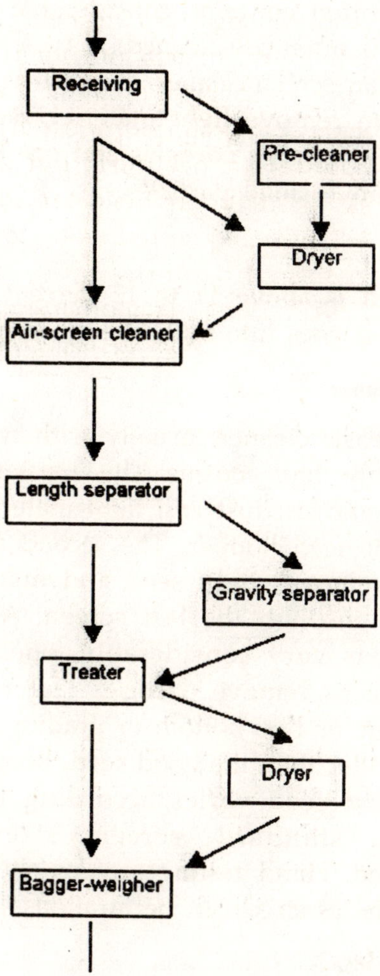

Figure 2. Flow diagrams of wheat seed processing

After the raw seed has been received and where necessary fumigated, processing operations (and sequence) are as outlined below (Figure 2).

Pre-cleaner

Wheat seed often contains considerable plant material trash, and it is often pre-cleaned. A typical pre-cleaner is similar to an air-screen cleaner, except that it has only one air channel to remove light material, one top scalping screen to remove large particles and one bottom grading screen to remove small particles.

Dryer

If wheat seed is above 11 to 12 percent moisture, it is dried before it goes into bulk storage or processing.

Air-screen Cleaner

This is the basic cleaner, usually with two air channels and, preferably, four screens. The first air channel (head aspiration) removes dust and light materials as the seed falls from the feed hopper. The second air channel (tail aspiration) removes light seed and materials after the seed passes through the last screen. Although screen configurations vary considerably, one or two top or scalping screens remove particles larger than the good seed, and one or two bottom or grading screens remove particles smaller than the good seed. Because the average size of wheat seed varies according to the growing conditions, standard screen sizes cannot be recommended. Hand testing screens should be used to determine the exact screen perforations.

Length Separator

A length separator is almost always used to clean wheat seed. By using the proper machine configuration, shorter or longer undesirable materials (such as broken grains, weed seeds, oat, barley, etc.) are removed. Broken grains and weed seeds, which are shorter than the good seed, are removed by using cylinders with smaller indents.

Larger impurities can be removed by using a cylinder with indents that lift all good seed, but contaminants (wild oats, oats or barley grains and unthreshed glumes) remain in the cylinder.

Gravity Separator

After the seed is cleaned by the air-screen cleaner and indented cylinder, it may be necessary to use a gravity separator. The gravity separator classifies a seed mixture mainly according to density or specific gravity. It can be used to remove unthreshed glumes and soil particles, which have similar sizes to wheat but different weights. Another application is the removal of weevil-infested grains from the seed lot and upgrading seed (in order to improve germination). Furthermore, wild oats and some barley may be removed from the wheat seed lots, but at the expense of substantial amounts of good seed and only after recycling the material a number of times on the gravity separator.

Treater

Wheat seed should, if necessary, be treated with the appropriate fungicide to protect the seed and seedling after planting. Insecticides are sometimes applied to protect seed in storage and in the soil. Treatments may be applied to protect the seedlings or adult plants against pathogens carried on or in the seed.

Dryer

In humid and hot climates, seeds may be sealed in vapourtight plastic bags to maintain viability over longer periods. In such cases, wheat seed moisture content must be below 9 percent, preferably not over 8.5 percent. Usually, a dehumidified, closed-circuit dryer is used after the seed treatment is applied.

Bagger-weigher

The final step is to weigh the proper amount of seed into the proper kind of bag. Wheat seed bags should be of a size that fits local farmer needs (seed rates and field size). During processing, strict attention should be paid to the cleanliness of the processing machines and any admixture should be avoided. Every processing plant should have a complete set of hand screens, a small air-screen cleaner and an indented cylinder to help determine the proper processing requirements. It is also essential to have an internal quality control laboratory attached to each seed plant with a small seed testing facility. This laboratory unit should constantly monitor the quality of the seed and the efficiency of processing operations.

Seed Treatment

Seed health is an important attribute of quality, and seed used for planting should be free from pests. Seed infection may lead to low germination, reduced field establishment, severe yield loss or a total crop failure. For example, severely infected wheat grains with Karnal bunt either fail to germinate or produce a greater percentage of abnormal seedlings.

In wheat, fungi (*Fusarium* spp., *Tilletia* spp., *Drechslera* spp., *Septoria* spp. and *Ustilago* spp.), bacteria (*Corynebacterium*, *Pseudomonas* and *Xanthomonas*) and nematodes (*Anguina tritici*) are the most important seed-borne diseases due to their worldwide distribution and losses they incur in crop production.

Chemical seed treatment is one of the efficient and economic plant protection practices and can be used to control both external and internal seed infection. It protects young seedlings or adult plants against attack from seed-borne, soil-borne or airborne pests. It disinfects seed from pathogen, checks spread of harmful organisms, promotes seedling establishment, maintains and improves seed quality or minimises yield losses. Selection of the proper chemical depends on the target organisms. A wide range of chemicals and equipment are now available for such purposes. Some recent literature gives detailed information on the management of bunts and smuts and bacterial seed-borne diseases of wheat.

Meisner *et al.* indicated that Vitavax 200 (Carboxin [37.5 percent] and Thiram [37.5 percent]) is an effective broad spectrum seed treatment fungicide, both for externally and internally seed-borne diseases of wheat. Moreover, pre-harvest foliar application of chemicals can also reduce the internally seed-borne fungi and can be combined with seed treatment to produce healthy seed. Sinclair cited that foliar spraying of wheat with benomyl, methyl benzimidazole carbamate or benomyl plus mancozeb reduced *F. graminearum,* whereas capatafol and mancozeb reduced *S. nodorum.*

Apart from disease control, seed treatment also has a positive effect on crop growth and yield. Ahmed reported that wheat seed treatment with systemic fungicides, such as Baytan, Raxil and Vitavax, significantly increased crop stand, grain yield and yield attributes. Meisner *et al.* reported a 10 percent increase in wheat yield due to seed treatment with Vitavax 200 against smuts.

Seed production in disease-free areas or under effective disease control and field inspection schemes is very important to obtain disease-free seed. Thus, understanding disease epidemiology, its transmission rate and economic threshold, combined with seed health testing, could help to define the need for seed treatment.

Storage

Seed is 'in storage' from the time it reaches physiological maturity on the parent plant until it is planted by the farmer. Germination is highest at physiological maturity, and viability then declines inexorably until the seed dies. Deterioration of seed viability cannot be reversed once it has occurred. Good storage cannot improve the quality of poor seed; therefore, only seed with high germination and high vigour should be put into storage. Storage conditions should then be as favourable as possible to maintain quality. Unfavourable conditions at any time during storage may reduce or destroy viability.

Seed should be harvested when it reaches harvest maturity, dried to a safe moisture content (if necessary), stored under favourable conditions and protected from damage and pests until it can be planted. Conditions that cause the loss of seed viability in storage include:

- immature or damaged seed cannot survive long storage periods. Seed should be harvested when properly matured;
- mechanical injury to seed during harvest or handling makes it more susceptible to deterioration in storage;
- seed should be properly dried before going into storage and protected from moisture and high relative humidity. Fungi (*Aspergillus* and *Penicillium*) cause damage to stored seed if seed moisture is high;
- high storage temperature has a damaging effect on seed. Stores should be designed so that low temperatures are maintained;

— some seed treatments cause seed to die if it is stored too long; therefore, seed should only be treated when it is certain that it will be sold for planting;

— rodents, mainly rats and mice, can be most destructive to seed. Effective rodent control (traps and poison) is essential in all seed stores. A complete programme of exclusion, sanitation and control should be used;

— insects should be controlled by a combination of insecticides and fumigants. Use safest fumigants (e.g. phostoxin) because some fumigants (e.g. methyl bromide) will reduce germination.

Keeping the seed as dry and cool as possible in clean stores is the best management practice. If seed is dry and cool, physiological processes, fungal activity and insect activity are low. Select a seed storage site that is cool and dry (low relative humidity). O'Dowd and Dobie described the design of open seed stores, and Ellis suggested a practical guideline in choosing alternative sites for short- and medium-term seed storage.

Clements discussed the problem associated with wheat storage under tropical conditions. Wheat seed is storable for medium to long periods if kept under safe storage conditions. For wheat, high seed moisture (above 11 to 12 percent) is the most damaging, and seed must be kept as dry as possible in storage. The response of wheat seed to high atmospheric humidity (RH) in storage varies with temperature. Clements reported that at 25°C and 75 percent RH the equilibrium moisture content for wheat is 15 percent, and at 90 percent RH this may increase to 19.7%. He also stated that the critical moisture content for wheat that increases the rate of respiration is 14.6 percent.

In general, stored wheat seed should be kept at moisture content levels below 12 percent and relative humidity should not exceed 50 to 60 percent.

Diekmann indicates that in cereals there are more than 20 different species of storage pests of which grain borer (*Rhyzopertha*) and weevils (*Sitophilus*) occur most frequently. Some insects, such as the khapra beetle (*Trogoderma granarium*), are quarantine pests in some countries. In India, it is reported that the rice weevil (*S. oryzae*), lesser grain borer (*R. dominica*), khapra beetle and flour beetle (*Tribolium castaneum*) are important seed storage pests of wheat. In Bangladesh, farmers traditionally use natural insecticides, such as neem (*Azadirachta indica*), biskatali (*Rumex obtusifolius*) and tobacco leaves, as insect repellents for farm-saved seed.

Seed Quality Control

The seed quality control and certification agency is responsible for ensuring variety purity, identity and other seed quality attributes, such as physical purity, germination and health. The official seed quality control agency carries out inspection in the field and tests the seed in the laboratory to confirm that the seed meets the national certification standards.

Variety purity and identity of seed is ensured through field inspection of growing crops. Land requirement, isolation distance and seed source are confirmed, and the presence of off-types, other varieties, other crops and seed-borne diseases are determined based on inspection of representative samples that are compared with the standards. Seed crops that meet minimum field standards are accepted for harvesting and processing.

After processing, the seed is tested for purity, germination, health, etc. These tests are carried out in a seed testing laboratory based on samples taken from cleaned seed lots. After a seed crop has been field inspected and laboratory tested, and standards have been met, labels can be attached to each sealed seed bag, and the seed is officially certified.

Table 4. Field and seed standards for wheat in Morocco

Standards	*Seed classes*			
	Pre-basic	*Basic*	*Certified 1*	*Certified2*
Field standards				
Rotation (*minimum, years*)	1	1	1	1
Isolation (*minimum,m*)[a]	10	10	10	10
Other varieties (*maximum, %*)	0.05	0.1	0.2	0.3
Other species (*maximum, number*)	1/15 000	1/10 000	1/5 000	1/2 000
Noxious weeds	-	-	-	-
Infected plants (*maximum, number*)[c]	1/10 000	1/5 000	1/2 000	1/1 000
Seed standards				
Pure seed (*minimum, %*)	99	99	98.5	98
Other varieties (*maximum, number/1 000 seeds*)	1	1	2	3
Weed seeds (*maximum, number/kg*)	6	8	20	30
Cereal species (*maximum, number/kg*)	1	2	12	15
Wild oats (*maximum, number/kg*)[d]	0	0	1	1
Noxious weeds (*maximum, number/kg*)	3	4	6	8
Infected seeds				
Loose smut and common bunt (*maximum, %*)	-	0.02	0.05	0.1
Insect infestation				
Live insects (*maximum, number*)	0	0	0	0
Insect damage (*maximum, %*)	0.1	0.1	0.1	0.1
Germination (*minimum, %*)	85	85	85	85
Moisture content (*maximum, %*)	14	14	14	14
Specific gravity (*minimum, hl/kg*)	79	79	79	79

Seed standards need to be rationalised, taking into consideration the relevance of the quality attribute, the experience of the seed growers and the level of agricultural development in the country. Excessively high standards that are copied from developed seed programmes may frustrate the growth of the seed industry and should be avoided. An example of field and seed standards for different classes of wheat seed used in Morocco is given in Table 4.

In many developed countries, there is a tendency to give more importance to the internal quality control by the companies that are producing and marketing the seed. In developing countries, seed quality control agencies often lack resources and are not able to carry out proper seed quality assessment. In view of this, the Food and Agriculture Organisation of the United Nations (FAO) proposal *Quality Declared Seed* is particularly relevant to developing countries because it aims at a system that requires fewer resources and still provides sufficient guarantee for the farmers.

Seed Marketing

Seed marketing is the final step in a seed programme; it takes the seed to farmers and gets them to buy it and plant it. Seed marketing is time-sensitive and sensitive to factors affecting rural marketing. Seed must reach the farmer at the right time, at the right place, at the right price, in the right amount and must be of the highest quality. Because seed marketing is sensitive to many factors, it is often considered a high-risk business. Seed marketing requires:

— convincing farmers that the seed quality is high and ensuring that only high-quality seed is sold;

- — convincing farmers that the seed quality means a benefit to them that is worth the extra cost they must pay for the seed. All possible means of promoting the seed must be used;
- — making the seed available in locations close to the target farmers. The seed must be readily available when the farmer needs it. It must be available in bag sizes that fit the farmers' needs.

Effective seed marketing is a specialised operation, as described by Gregg and van Gastel. Wheat seed marketing has the same requirements, problems and needs as the marketing of other crop seed. Farmers' variety and seed replacement is influenced by their perception of the yield gain of a new variety, the yield loss of an old variety and the risk in changing the variety. In wheat, decline in variety purity and yield from retained seed is expected to be low. An estimated yield reduction of 0 percent in dry areas of Australia, 0.25 percent in Pakistan and 1.6 percent in Nepal from retained seed is reported. As a result, farmers can save their own seed for planting for quite a long period of time once they have purchased seed of the new variety.

Few farmers are willing to spend more money to purchase improved seed of wheat. On average, wheat farmers typically purchase a relatively small amount of improved seed.

6

Fruits

Fruits are so varied in form and development, that it is difficult to devise a classification scheme that includes all known fruits., It will also be seen that many common terms for seeds and fruit are incorrectly applied, a fact that complicates understanding of the terminology. Seeds are ripened ovules; fruits are the ripened ovaries or carpels that contain the seeds. To these two basic definitions can be added the clarification that in botanical terminology, a nut is a type of fruit and not another term for seed.

There are three basic types of fruits:

— Simple fruit

— Aggregate fruit

— Multiple fruit

Simple Fruit

Simple fruits can be either dry or fleshy and result from the ripening of a simple or compound ovary with only one pistil. Dry fruits may be either dehiscent (opening to discharge seeds), or indehiscent. Types of dry, simple fruits (with examples) are:

— achene - (buttercup)
— capsule - (Brazil nut)
— caryopsis - (wheat)
— fibrous drupe - (coconut, —walnut)
— follicle - (milkweed)
— legume - (pea, bean, peanut)
— loment
— nut - (hazelnut, beech, oak acorn)
— samara - (elm, ash, maple key)
— schizocarp - (carrot)
— silique - (radish)
— silicle - (shepherd's purse)
— utricle - (beet)

Fruits in which part or all of the pericarp (fruit wall) is fleshy at maturity are simple fleshy fruits. Types of fleshy, simple fruits (with examples) are:

— berry - (tomato, avocado)
— stone fruit drupe (plum, cherry, peach, olive)
— false berry - accessory fruits (banana, cranberry)
— pome - accessory fruits (apple, pear, rosehip)

Aggregate Fruits

Several types of aggregate fruits exist:

Achenacetum: Cluster of achenia; e.g., the strawberry (*Fragaria vesca).*

Baccacetum *or etaerio.* Aggregate of berries; e.g., Annonaceae (*Asimina triloba, Cananga odorata, Uvaria*). The berries can be aggregate and syncarpic as in *Annona reticulata, A. muricata, A. pittieri* and other species.

Drupacetum: Aggregate of druplets; e.g., *Bursera simaruba* (Burseraceae).

Folliacetum: Aggregate of follicles; e.g., Annonaceae (Anaxagorea crassipetala, A. phaeocarpa, Xylopia aromatica, X. frutescens, Cymbopetalum costaricense, C. torulosum, Guatteria), Illiciaceae (Illicium verum, I. anisatum), Apocynaceae (Aspidosperma, Prestonia surinamensis, Stemmadenia, Tabernaemontana), *Magnoliaceae (*Magnolia poasana, Talauma gloriensis*), and Sterculiaceae (*Helicteres guazumaefolia, Sterculia*).* Desmopsis bibracteata *(Annonaceae) has aggregate follicles with constrictions between successive seeds, similar to those found in loments.*

Samaracetum: Aggregate of samaras; e.g., Aceraceae (Acer pseudoplatanus), Magnoliaceae (Liriodendron tulipifera *L.), Sapindaceae (*Thouinidium dodecandrum*), and Tiliaceae* (Goethalsia meiantha).

Multiple Fruits

Multiple fruits are found along a single axis and are usually coalescent. The most common types follow:

Bibacca: Double fused berry; e.g., *Lonicera.*

Sorosis: Fruits usually coalescent on a central axis; they derive from the ovaries of several flowers; e.g., Moraceae (*Artocarpus altilis*).

Syconium: Syncarp with many achenia in the inner wall of a hollow receptacle; e.g., *Ficus.*

Gymnosperm Fruit

Fertilisation stimulates the growth of young gynostrobiles which in species such as *Pinus* are more than 1 year old. Many genera have woody gynostrobiles (*Pinus, Picea,*

Pseudotsuga); others have fused scales forming a berry-like structure around the seeds (*Juniperus*). In some species, such as *Taxus* and *Torreya*, the seeds develop inside fleshy arils. In *Podocarpus* and *Prumnopitys*, the ovuliferous scale forms a fleshy covering called epimatium, which surrounds the seed. This structure can be partially or totally fused to the ovule integument and even to the bract subtending the ovuliferous scale.

The pre- and postfertilisation development of the strobile shows many facets similar to those of angiosperm fruits. Parameters such as size and moisture content show daily fluctuations as a consequence of variations in the moisture level, increasing dry weight, high respiratory rates, and storage of carbohydrates and minerals. However, in the last stages of maturation the moisture content decreases. Weight may be lower, respiration is lower, and nutrients are mobilised to the seed. The metabolic activity declines with strobile dehydration and the scales become open.

Seed

The seed is the site of a partial development of the sporophyte (*embryo*) and the linkage between successive generations. It is a critical intermediate stage in the life cycle of angiosperms and gymnosperms, which guarantees the propagation and survival of the species.

Seeds are dynamic and tridimensional entities, and their morphology is the result of the physiological and environmental processes involved in their development. The physical configuration (size, shape, distribution, and structure of tissues and organs) influences, at different ontogenetic stages, the nature and efficiency of the functional activities.

Angiosperm Seed

Seed development

The ovule, whose embryo sac was fertilised twice, is the starting point for seed development. In some cases, the process is long and complicated; in others, it is short and simple.

In seed development, three functional phases can be identified:

1) Cell divisions produce the tissues that will form the seedcoat, the endosperm, and the embryo (early embryogenesis); this stage is characterised by a fast increase in fresh weight.
2) Ontogenetic changes guarantee the success of the offspring as an independent unit, through the storage of reserves. They lead to an increase in dry weight.
3) Maturation drying leads to a stage of metabolic quiescence, interpolated between the end of seed development and the beginning of germination. In this stage, fresh weight decreases. Many studies suggest that this period of dehydration is important to the transition of activities from seed development (especially embryo) to germination-seedling development.

Maturation drying guarantees the permanent inactivation of seed metabolism during seed dispersal and the period before germination. The reduction of water content in the tissues, the impermeability of the seedcoat, and the presence of inhibitors cause seed inactivity. The influence of these factors varies from one species to another, but many seeds do not germinate if they are removed from the maternal plant before this stage. However, a precocious germination may be induced through a dessication-rehydration period. The maturation drying stops seed development and irreversibly stunts the germinationseedling development. A dramatic reduction in the synthesis of storage proteins occurs. Synthesis of reserves stops and synthesis of proteins related to germination-seedling development is introduced (e.g., enzymes related to reserve mobilisation). The change in genetic expression seems to act at the transcription and post-transcription (mRNA) levels. The drying suppresses the production of messages to synthesise proteins for development. When the seeds rehydrate, the messages to synthesise proteins used in development and growth increase.

Growth regulators present in the tissues of developing seeds (indol acetic acid, IAA), gibberellins (GAs), cytokinins, and abscisic acid (ABA) seem to be involved in several processes. These processes are development of the seed (growth: cell division and elongation) and cellular differentiation (qualitative differences between cells, tissues, and organs), including the arrest of growth prior to seed germination; accumulation of the storage reserves; development of the extraseminal tissues (growth and cell differentiation); and several physiological effects on tissues and organs close to the developing fruit.

The biologically active gibberellins show high concentrations during the growth (mainly cell division) of the embryo and the endosperm. Abscissic acid plays an important role in the development and maturation of seeds. It is associated more with the reduction of embryo growth than with its promotion; in many cases, it can be present in the normal embryogenesis but not in the germination and longitudinal extension of the embryo axis. Abscissic acid concentration also appears to influence the deposition of proteins and other reserves and, in many species, the highest content of ABA is coincident with the highest rate of reserve synthesis. It also induces dehydrine expression (LEA), which is the accumulation of proteins during the last stages of seed maturation (late embryogenesis). However, the entire role of ABA is not yet well comprehended. Dehydrine expression finishes when germination starts. Water loss during the process of maturation is common to many seeds, which may diminish to less than 5 to 10 percent of their fresh weight.

In temperate zones, and especially in the tropics, many seeds do not undergo drying, do not experience reduced cellular metabolism, and do not exhibit a clear end to seed development. During fruit dehiscence and dispersal, seed development is followed by germination-seed development without interruption. In some species, the seedling develops when the seed is still inside the fruit and is attached to the parent tree. These seeds are called *viviparous,* and it has been demonstrated experimentally that the inhibition of their development requires high concentrations of ABA. Mangroves growing on protected tropical coasts are the best-known examples of viviparous seeds.

Common on the American coasts are *Rhizophora racemosa* (Rhizophoraceae), *Pelliciera rhizophorae* (Theaceae), *Avicennia germinans* (L.) L., *A. bicolor* (Verbenaceae), and *Laguncularia racemosa* (L.) C.F. Gaertna. (Combretaceae). *Inga paterno* Harms (Fabaceae-Mimosoideae) also has a viviparous seed. *Cojoba arborea* and *C. costaricensis* (Fabaceae-Mimosoideae) show an incipient viviparity. The overgrown seeds described by Corner, which have a continuous embryo development limited by a hard and indehiscent pericarp or a late dehiscence pericarp (*Dipteryx panamensis, D. odorata,*

Prioria copaifera), seem to be an intermediate type. In these seeds, the cotyledons fold and become disfigured because space is inadequate. The plumule, which is well developed, usually has an incipient epicotyl and several foliar primordia in different stages of development.

Embryogenesis

Embryo development includes two fundamental processes: the establishment of a precise spatial organisation of the cells derived from the zygote (pattern of formation), and the generation of cell diversity inside the developing embryo (cytodifferentiation). These processes are coordinated to develop a recognisable morphological structure, regulated by the embryogenic pattern of the species. Although in plants most organogenesis occurs in the postembryonic stage, the embryonic pattern reveals coordinated growth and development during the process of development.

The pattern of embryogenic formation has three levels: pattern of cellular organisation, pattern of protein accumulation, and pattern of gene expression.

The unicellular zygote evolves into a multicellular embryo with differentiated organs and the potentialities of the adult plant. This occurs through a programmed sequence of events (pattern of embryogenesis). In angiosperms, the embryo is located at the micropylar end of the embryo sac. Its basal end is attached to the embryo sac wall and its apical end faces the central cell. After fertilisation the zygotic cell polarises (establishes structural and physiological differences). The endoplasmic reticulum and the cellular organelles reorganise around the nucleus.

During mitosis, the polarised zygote divides, forming two daughter cells that have cytoplasmic elements differentially distributed. This fact has a strong effect on the embryogenesis. In most angiosperms the zygote divides transversely, producing a chalazal or apical cell (ac) and a basal cell (bc); in several exceptional cases, cell division is longitudinal or oblique as in Piperaceae and Dipsacaceae. The cells derived from the apical cell contribute to form most of the proembryo and embryo (usually the plumule and the cotyledons); the basal cell and its derivatives commonly form the hypocotyl and the suspensor.

The variations in cell wall orientation and the sequence of cell divisions in the embryo are the basis for classification of embryonic types. The separation of types, except the piperad, is based on the destiny of the proembryo apical cell during the second and subsequent cell divisions, and in the contribution of the basal cell to the embryo organs. In the onagrad type, a vertical division in the apical cell (ac) and a transversal division in the basal cell (bc) result in a tetrad of cells forming a "T". The ac and its derivatives form most of the embryo (cotyledonar zone, apical meristem, hypocotylar zone). The bc forms the radical cortex, the root cap, and the suspensor. The proembryo in the asterad type is formed through a vertical division in ac and a transverse division in bc.

In some taxa, the ac divides obliquely and when the quadrant is formed, one of the cells forms the epiphysis. Through further growth, the epiphysis produces the epicotyl. The remaining cells in the quadrant develop into the cotyledons. In this type, bc contributes to form a considerable portion of the embryo (hypocotylar region, radical cortex, root cap, suspensor). The solanad type forms a linear tetrad of cells. The two cells derived form ac from most of the embryo; bc does not take part in embryo organisation but forms the root cap and the suspensor.

The chenopodiad type also has a linear tetrad; ac produces the cotyledonar region, the apical meristem, and a hypocotyl half, while bc provides the other hypocotyl half, the radical cortex, the root cap, and the suspensor. The caryophyllad type differs from the other types. Here, the bc does not divide and the ac originates the embryo. The piperad type includes those cases in which the zygote divides in a longitudinal plane.

The time and the fixation mechanism of the polar axis in the zygote are very important in morphogenesis. The establishment of the polarity, and the zygote and proembryo histodifferentiation occur in the absence of cell movement due to the rate and planes of cell division and subsequent cellular elongation. As a result, the regions giving rise to the apical and radical meristems are formed. These changes mark the beginning of mRNA synthesis in the zygote during early embryogenesis.

During the zygotic, proembryonic and embryonic stages, coordination exists at the biochemical level. A specific pattern also exists for the synthesis of proteins, lipids, and carbohydrates in the embryo and other seed structures. The three types of protein involved in this synthesis are as follows: enzymes involved in reserve mobilisation, structural proteins associated with the membranes and the ribosomes, and storage proteins in which the LEA or dehydrines are critical.

The activation and expression of genes during embryogenesis, especially those regulating specific aspects of development, are the subject of increasing research. Lindsey and Topping identified a gene in *Arabidopsis* that expresses in the basal region of the proembryo when the latter is in the heart stage, and in the radical meristem when in the adult stage. By studying mutants, researchers have proven that numerous loci control the diverse aspects of embryogenesis. Processes such as cell division and morphogenesis per se, and the establishment of polarity, embryo, and endosperm shape are determined by genes that can be recognised easily.

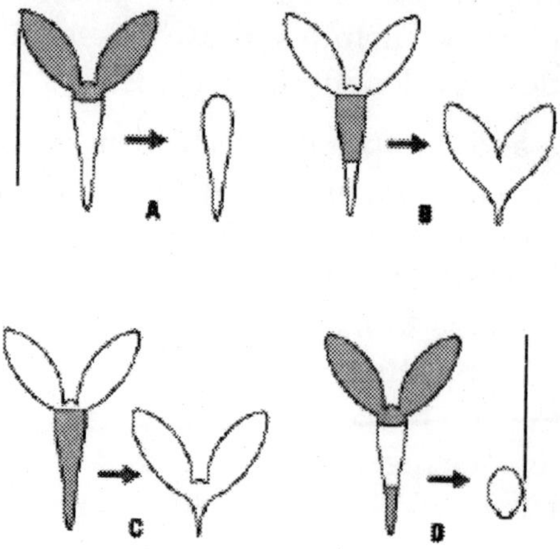

Figure 1. Origin and final form of apical-basal pattern mutants.

In the mutant a, the apical cell does not form the plumule, and the cotyledons resemble the embryo of some Lecythidaceae such as *Lecythis ampla*. Perhaps the embryology of *L. ampla* will reveal similarities to that found in mutant *a*.

Embryo Differentiation

The tridimensional growth of the embryo begins with the formation of vertical walls in the derivatives of the apical cell. Further divisions in several planes produce a globular embryo. Cell differentiation begins in this stage with the establishment of the cotyledonar and epicotylar loci in the shoot pole; the hypocotyledonar region and the radical pole with the hypophysis are also distinguished. Loci organisations are not strictly related to cell lineage or row systems. The cotyledonar, epicotylar, hypocotyledonar, and hypophysial regions are identified after the growth and differentiation of the incipient organ. The hypocotyledonar region contributes to the embryo axis formation through cell division and elongation. This phase of differential growth in the embryo leads to the establishment of the meristems and accentuates the differences between shoot and root.

Embryo histogenesis in seeds without maturation drying is similar in its early stages to that of seeds with maturation drying; however, accentuated variations appear in late embryogenesis. Many seeds without maturation drying are large, with very well developed embryos such as those of *Aesculus hippocastanum*, *Dipterocarpus*, *Quercus robur*, *Hevea brasiliensis*, *Calophyllum brasiliense*, *Mangifera indica*, and *Sclerocarya birrea*. Other seeds are large but have a small, rudimentary embryo such as those of *Hyeronima alchorneoides*. All these seeds increase in dry weight until fruit dehiscence, with slight or no loss in fresh weight; however, a decreasing loss in water content is characteristic of several seeds without maturation drying. Embryo growth may continue (increasing in dry weight) after dehiscence in the absence of enough water to promote germination; however, in the tropical forest, with very high rain regimes, the seeds or fruits (diaspores) fall down on very humid soils, sometimes inundated, and continue hydrating.

Dicotyledons and monocotyledons have similar types of embryonic development. During early embryogenesis–the stages of quadrant and octant and the formation of the globular embryo–the cells in both groups are similar in lineage and configuration. The fundamental differences arise during differentiation of the globular embryo.

The organisation of the shoot apex follows different morphogenetic patterns. In dicotyledons, the axial cells forming the epiphysis grow slower than the circumaxial cells. In monocotyledons, one-half of the terminal cell and its derivatives grow slowly while the remaining half, which forms the cotyledonar locus, grows rapidly. The apparent lateral position of the shoot apex in late stages is due to the rapid growth of the single cotyledon in monocotyledons. The epicotyl and the cotyledon of monocotyledons develop from the same layer of terminal cells.

The two loci differentiate through a vertical wall in the terminal layer. The differences found in the terminal region are conspicuous during the quadrant stage. In the dicotyledons, two cells opposite the terminal quadrant form the cotyledons; in the monocotyledons the number of cells forming the cotyledon in the quadrant stage is variable. Despite these variations, the changes occurring in the different ontogenetic stages result in an organised morphogenetic system from which the future tree evolves.

Suspensor in Embryogenisis

The suspensor plays an active and dynamic role during embryogenesis. The suspensor has a stable cell number and its rate of growth is higher during the early stages of embryogeny. In many species it has transfer cells, suggesting an active role in the absorption and exchange of nutrients for embryo growth. During the late stages of the embryogeny, the suspensor degenerates and seems to be digested by the embryo.

The suspensor seems to accomplish two functions: (1) nutrient absorption from the surrounding somatic tissues and transportation to the developing embryo; and (2) a source of nutrients and growth regulators for the developing embryo. In vitro, gibberellic acid can partially replace the suspensor; therefore, the suspensor must be a source providing this growth regulator to the developing embryo. The suspensor also contains auxins and cytokinins; the concentrations of these substances have the same fluctuation pattern as those of gibberellic acid.

The embryo of some taxa (*Tilia*) do not have a suspensor; others have a reduced structure (*Euphorbia*, *Ruta*). Usually, the taxa with haustorial endosperm do not have a suspensor or have one that is reduced and ephemeral. Families such as Rubiaceae have a massive and haustorial suspensor. In Fabaceae, the suspensor may be missing or may form a massive structure, filamentous or tubercular.

Embryogenic Deviations

In some parasite and saprophyte taxa, as well as in some forest trees [Olacaceae (*Minquartia guianensis*), Myristicaceae (*Virola* koschnyi, V. sebifera, Otoba novogranatensis), Icacinaceae (Calatola *costaricensis*), Piperaceae (*Piper*, *Peperomia*)] minute embryos are found. Frequently they are also rudimentary. Other embryos have deviated since their origin. Some deviations are formed by apomixis (asexual process); they may derive from the unfertilised egg cell (haploid parthenogenesis) or from another cell from the gametophyte (haploid apogamy). In many cases, there is no meiosis and a diploid gametophyte is formed, leading to diploid parthenogenesis or diploid apogamy.

Endosperm Development

The fusion of the sperm nucleus with the central cell nuclei forms the primary endosperm cell. Through successive mitosis, this cell forms the endosperm. This tissue does not have a significant role at the proembryo stage, but it is important to embryo nutrition during embryo development and seed germination. In most dicotyledons the endosperm is reabsorbed during seed maturation and storage proteins accumulate in the tissues of the embryo. During seed development, the interactions between endosperm and embryo are essential to insure reproductive success. The formation of the endosperm, its reabsorption, and the transference of reserves to the embryo are genetically established.

Three types of endosperm are recognised: nuclear, cellular, and helobial. Many free nuclear divisions before cell wall formation characterise the nuclear type; its formation varies from one species to another. The endosperm is consumed before partial or total wall formation occurs in many species. The seeds of other species, such as *Virola koschnyi, V. sebifera, V. surinamensis,* and *Compsoneura sprucei,* have nuclear endosperm when they are dispersed. Several species have haustorial endosperm.

In the cellular endosperm, each cell division is followed by cell wall formation; this type of endosperm is frequently haustorial. The helobial endosperm has a different development: the first endosperm cell divides and gives rise to a pair of unequal chambers; usually, the micropylar chamber is bigger. It undergoes free nuclear divisions before cell wall formation; in contrast, the nucleus of the chalazal chamber does not divide or undergo division before giving rise to a coenocyte. Sometimes cell wall deposition takes place later. The helobial endosperm is typical of the monocotyledons.

Mature Seed

The mature seed generally has a seedcoat (product of one or both ovule integuments), an endosperm, and an embryo. Some mature seeds retain a remnant of nucellar tissue called a perisperm. The degree to which these structures continue their development, are reduced or reabsorbed, or disappear during the late stages of seed development leads to distinct structural patterns associated with physiological differences.

Seedcoat

In bitegmic seeds the seedcoat has a testa (former outer integument) and a tegmen (inner integument). Each integument has its own opening at the seed's distal end; the opening in the outer integument is the exostome while that of the inner integument is the endostome. The exostome and the endostome form the micropyle. Unitegmic seeds have only one opening (micropyle) and the seedcoat is called the testa.

The seedcoat has a funicular scar called the hilum, indicating the point at which the funiculus separates from the seed. This zone has a different structure and often includes the micropyle. Frequently, a longitudinal ridge (raphe) formed by differential growth of the funiculus is found near the hilar zone. It forms as a result of a curvature of about 80 degrees that occurs during the ontogeny of the anatropous ovules. The hilar zone may extend and be a significant part of the seedcoat as observed in the Sapotaceae (Manilkara chicle, M. zapota (L.) P., Micropholis crotonioides, Pouteria congestifolia, P. viridis), or may combine its extension with an extension of the chalazal zone (pachychalaza) as seen in the Meliaceae (*Carapa, Guarea*), and Hippocastanaceae (Aesculus, Billia columbiana, B. hippocastanum).

Other seeds have only a chalazal extension (*Cupania glabra*). A funicular protuberance close to the chalaza is found in many seeds (numerous Cactaceae in the Subfamily Cereaneae). The vascular bundle supplying the ovule follows its course along the raphe—a characteristic easily observed in the seeds of Fabaceae-Papilionoideae (*Erythrina*).

The seedcoat is formed not only by the integument tissue(s) but also by the chalazal and raphal tissues. When the outer layer of the outer integument differentiates as a layer for mechanical protection (exotesta), the outer layer of the raphe and the chalaza undergo an equivalent differentiation. The same happens with the formation of the mesotesta (middle layers) and the endotesta (inner layer of the outer integument). In small seeds, the integument cells undergo mitosis sporadically after fertilisation, and their elongation and differentiation form the seedcoat.

Many large seeds have complex seedcoats as a result of many anticlinal and periclinal divisions in the integument cells. When periclinal divisions are dominant and more cell layers are formed, the integument is multiplicative (Magnoliaceae, Myristicaceae, Annonaceae, Clusiaceae, Fabaceae, Bombacaceae). If the seedcoat extends primarily by anticlinal cell division and does not develop more cell layers, the integument is nonmultiplicative. Frequently, multiplicative integuments form seedcoats with many layers. The periclinal divisions may be diffused or localised in the outer integument.

The testa can be soft-textured (Cojoba arborea, C. costaricensis), hard (Enterolobium *cyclocarpum, Samanea saman, Tamarindus indica*), or fleshy (*Inga, Guarea, Punica granatum, Carica papaya, Magnolia*). The fleshy testa is the sarcotesta. In general, the outer integument is the fleshy and juicy part; however, in *Magnolia* and other related species as well as in *Punica granatum*, the sarcotesta originates only from the mesotesta. In Connaraceae, Meliaceae, and Sapindaceae the fleshy pachychalasa is similar to the aril. It may have lipids, sugars, and attractive color.

The presence of a multiplicative mesotesta is common; these layers may form an aerenchyma (Burseraceae, Bombacaceae, Meliaceae, Simaroubaceae, Fabaceae). Usually the endotesta divides anticlinally. An endotesta with multiple layers is common in the zoochorous seeds. In most species, the inner integument is thin and slightly specialised; it collapses in early stages and is then partially or totally reabsorbed. In exceptional cases, it forms multiplicative middle layers that produce the aerenchyma (e.g., many Euphorbiaceae), although in several families those layers disintegrate during seed maturation.

The seedcoat may have cell layers or cell groups with tannins, oils, crystals, mucilage, cork, sclerenchyma, or collenchyma cells. The parenchyma may serve as storage tissue, as chlorenchyma, or as aerenchyma. The tannins form in vacuoles, and are formed by polyphenoles and substances of hard metabolic transformation. The presence of tannins serves as a requisite for alkaloid deposit and probably protects the plant against herbivore predators (insects), pathogens (bacteria, fungi), and light. They also increase seed hardness and provide color.

The crystals are variable in shape and are formed by calcium oxalate, calcium carbonate, or silica. The mucilage cells are found primarily in the epidermal cells of the exotesta. They form by hydration of substances in the secondary walls of the cells. Some species in Sterculiaceae have mucilage sacs in the testa and tegmen. In Ebenaceae, these sacs are restricted to the tegmen. Mucilage secretion (myxospermy) appears to help seeds adhere to animals and fix in the soil. Other proposed functions of myxospermy include water retention, germination regulation, oxygen barrier, storage substance, and covering lens.

Sclerenchyma and collenchyma cells provide mechanical strength. In some seeds, cell walls impregnated with suberin serve as a barrier in the chalazal zone and seal the testal or tegmic layers. These cell walls are also found in the integuments of the hydrochorous seeds and in mucilaginous epidermal cells. The sclerenchymatic cells (fibers or sclereids) provide hardness and rigidity to the seedcoat. Additionally, the lignin protects against herbivore and pathogen attacks. The more typical macrosclereids are the Malpighi cells of the exotesta in Fabaceae.

These cells are elongated radially, have a linea lucida (light line), and in early stages deposit a wall of irregular thickness, which is sometimes lignified. The apical region shows suberised incrustations reinforcing the walls and influencing water permeability. The linea lucida is not involved in cell wall permeability and shows a high density of cellulose microfibrils without interfibrillar spaces. The seeds of Rhamnaceae and Elaeagnaceae have a linea lucida in the external epidermis of the testa. The structure is also present in the external epidermis of the tegment.

When the sclerenchymatous layer forms in the external epidermis of the testa, the seedcoat is called exotestal. The mesotestal seeds have sclerenchyma in the testal middle layers; the endotestal, in the epidermis of the testa. If the sclerenchyma differentiates in the tegmen, the seeds are exotegmic, mesotegmic, and endotegmic respectively. In families with dry indehiscent fruits or drupaceous fruits (e.g., Anacardiaceae), there is a marked tendency toward seeds with a slightly differentiated testa, sometimes lacking mechanical layers. In extreme cases, the mature seed lacks a seedcoat, as in parasites, some Apocynaceae, and Rubiaceae.

The seedcoat has an external cuticle on the external epidermis of the testa, a middle cuticle between testa and tegmen (if both exist), and an inner cuticle between the epidermis of the inner integument and the nucellus. The cuticles contribute to the impermeability of the seedcoat (to water and gases) and may influence the metabolism and growth of the embryo. The external surface of the seedcoat has characteristics of taxonomic value. Its morphology sometimes shows the influence exercised by the endocarp of the fruit or by arilar structures (e.g., Myristicaceae). Characteristics such as cell distribution, form, surface (conic, papillate, reticulate, striate, micropapillate, or hairy external wall), and epicuticular waxes (scarce) define the external surface of the seed. In many species, the seeds have stomata localised in the external epidermis of the testa.

Some families, such as Myristicaceae, have seeds with an irregular tegmen producing invaginations inwards; these invaginations protrude into the endosperm. This type of endosperm is ruminate. The seeds with a network of lobules in the endosperm are labyrinthine. The lobulation can be the result of the invagination of the tegmen and the folding of the embryo cotyledons. The testa has special structures such as the caruncle, strophiole, and aril. The caruncle is a fleshy protuberance or swelling arising in the exostome (*Ricinus, Euphorbia*). Some caruncles (elaiosomes) are oily, or rich in ricinoleic acid (e.g., *Turnera*). These caruncles are white or yellow, fleshy, and edible and are detached, eaten, and dispersed by ants. The strophiole is an outgrowth of the raphe. The aril is a funicular or hilar outgrowth and may cover the whole testa (complete) or part of it (incomplete; e.g., *Stemmadenia, Lecythis ampla, Phitecellobium dulce*). Some arils are reticulate or have frimbriate projections (frimbriate arils; e.g., *Myristica fragrans, Virola, Otoba*). The aril develops after fertilisation. When complete, the aril may cover the micropyle. Morphologically the caruncle is considered an exostomic aril and the strophiole a raphal aril.

Many seeds of Fabaceae-Mimosoideae (*Albizia niopoides* (Spruce ex Benth.) Burkart, *Entada, Enterolobium cyclocarpum, Pseudosamanea guachapele* (Kunth Harms, *Stryphnodendron microstachyum, Samanea saman*) are laterally compressed and have a special mark on each lateral surface called a pleurogram and a fissure (linea sutura or linea fissura) delimiting the pleurogram. The linea fissura is parallel to the raphe-antiraphe line and breaks to the hilar end. The linea fissúra seems to be associated with the gradual dehydration of the seed during maturation and storage, and the slow imbibition of the seed during germination.

The wings or hairs found on many seeds seem to be devices for anemochorous dispersal. Some winged seeds (*Vochysia, Qualea, Cedrela, Swietenia, Bernoullia flammea*) have one to several testal extensions that form wings; in many seeds the wing is circular or oval in outline (e.g., *Tabebuia rosea, Jacaranda copaia*). Other seeds have a tuft of trichomes at one end (*Macrohasseltia macroterantha*) or hairs distributed at random. The crowned seeds have a crown of hairs at one end, and the crestate seeds have one or several ridged excrescences. There are also punctate seeds covered by small excrescences or depressions, round umbonate seeds with a central or lateral umbo, and warty seeds with wart-like foldings.

Vascular System of the Seed

The vascular system of the ovule is formed by procambial or partially differentiated vascular strands. The differentiation of the vascular system occurs during the development of the seed. The vascular system supplying the seed is formed by placental, raphal, and funicular vascular bundles. These bundles transport nutrients to the developing seed. The larger seeds have a more extensive and well-developed vascular system; the small seeds have partially differentiated or no vascular bundles. The large seeds usually have a massive or compound raphal vascular bundle, well differentiated, which ends at the chalaza. Some seeds have additional vascular bundles extending from the chalaza to the outer integument; these are called postchalazal vascular bundles.

Some seeds are encircled by a postchalazal vascular bundle along the median plane (Annonaceae, Polygalaceae) and others have a chalazal vascular plexus and a network of vascular bundles. Through intercalar growth, this plexus may replace the seedcoat locally (e.g., *Guarea*). Vascular ramifications in the tegmen are common. In the seeds of species such as the Myristicaceae, the vascular branches enter the tegmen invaginations, increasing the contact with the endosperm. In *Virola koschnyi*, *Otoba novogranatensis*, and *Compsoneura sprucei* the vascular branches found in the tegmen invaginations have transfer cells.

Storage of Reserves

Seeds contain many substances. Some are typical to cells and tissues and others contribute to nourishing the embryo (before and after germination) and the developing seedling (in its early stages). Seeds store energy in the form of lipids, carbohydrates, and proteins to fulfill the needs of the seed during germination. They are a source of precursors for carbon skeletons and a source of energy when assembling precursors. The seeds are perispermous if the reserves are stored in the perisperm, endospermous or albuminous if the nutrients accumulate in the endosperm, cotyledospermous if the thick and fleshy cotyledons store the reserves, hypocotylospermous or macropodial when the hypocotyl is the storage organ, and chalazospermous if the nutrients are stored in the chalaza.

The cell walls of different seed tissues may also store nutrients. Lipids, which appear as lipidic bodies in the endosperm and the embryo, are a greater source of nourishment than the carbohydrates. The amount of lipids in a seed varies: 30 percent for the sunflower *Helianthus annuus* and 50 percent for *Ricinus communis, Zea mays,* and *Arachis hypogaea*. In the coconut (*Cocos nucifera*) and the African oil palm (*Elaeis guianeensis*) the lipid content is higher. The seeds without maturation drying are rich in lipids. For example, the lipid content in fresh seeds of *Virola koschnyi* is approximately 41 percent, in *Calophyllum brasiliense* 38 to 39 percent, in *Minquartia guia-nensis* 37 to 38 percent, and in *Lecythis ampla* 40 percent. The dried cotyledons of the embryo of *Carapa guianensis* have 65 to 70 percent of unsaturated lipids in the storage parenchyma; the endosperm of *Otoba novogranatensis* has about 69 percent of lipids (dry weight).

Carbohydrates are stored as starch or in thick cell walls rich in hemicelluloses. Cereals contain 70 to 80 percent starch, legumes ± 50 percent. The predominant types of carbohydrates stored in cell walls are mannans, xyloglucans, and galactans. The three types of mannans–pure mannans, glucomannans, and galactomannans–are restricted to the endosperm cell walls. The pure mannans are found in the date (*Phoenix dactylifera*), the ivory nut or tagua (*Phytelephas macrocarpa*), the coffee bean (*Coffea arabica*), and *Carum carvi*.

The galactomannans are found in the endosperm of the legume seeds. The mannans and glucomannans are crystalline and insoluble; they confer an extraordinary hardness to the endosperm. The xyloglucans are amiloid and are found in the endosperm or the embryo of approximately 2,600 species, among them *Tamarindus indica* and *Annona muricata*. The galactans are stored in the cotyledons of the embryo of numerous legumes.

Nearly all seeds contain proteins as reserve. Proteins supply the nitrogen needed by the plant in early stages of development. The storage proteins are found as protein bodies (aleurone). These grains are the primary source of proteins and minerals, and in addition to the homogeneous protein matrix they contain crystals of proteins and calcium oxalate. Several cations (K, Mg, Ca, Fe, Ba, Mn) are also found as globoid crystals. The protein bodies are found in the embryo as the endosperm or are restricted to a specialised layer (Poaceae, Fabaceae). This layer is active only during germination. Protein reserves are necessary to synthesise the enzymes involved in starch digestion. Nuts are about 40 percent proteins. The following species have seeds with a high protein content: Aesculus hippocastanum, Billia columbiana, Bertholletia excelsa, Lecythis ampla, Enterolobium cyclocarpum, *Leucaena leucocephala* (Lam.) de Wit, and *Entada scandens.*

The seeds with maturation drying accumulate disaccharides, such as the saccharose and oligosaccharides, in the form of stachyose and raffinose. Some propose that these sugars are associated with tolerance to desiccation; however, some seeds sensitive to desiccation also accumulate sugars and saccharose (*Avicennia marina*) or saccharose and raffinose (e.g, *Quercus robur*).

7

Pollination

Climatic factors such as wind, rain, moisture, temperaturc, light intensity, and spectral quality are the stimuli that determine the floral anthesis and the activity of the pollination vectors. The constancy or continuous change of these factors, the graduation or abruptness of any change, and the diurnal or seasonal periodicity of changes can delay or impede pollination. The covering of the pollen and the stigma are involved in the intercellular recognition of signals and stimuli. The first interaction of pollen and stigma is the capture of pollen by the stigmatic surface, usually 30 to 60 seconds after the initial contact. The epicuticular waxes of the stigmatic surface and the unsteady oils of the exine are placed strategically to function as the initial lipophilic intercellular bond between the matrices of the exine (pollen) and the cutin (stigma). Selective capture of compatible pollen is not apparent in this initial interaction.

The second cellular interaction is slower and takes place between pollen and stigma compatibles. When the pollen and the papillae or stigmatic surface come into contact, they form rigid convex surfaces tangentially between them; the macromolecular distances are long and varied. The hydrolysis and polysterification of the substances (oils and waxes) that made the initial contact connect the pollen exine with the cutin matrix. The enzymes of the cuticle (cutinase, estearase) catalyse the process.

Usually 15 to 30 minutes after pollination, a tubular connection is established between the pollen and the stigmatic surface. This connection has a diameter double the size of the orifice left by the pollen tube when it crossed the cuticle. The pollen tube develops between the connection, tube, or ocreatine and goes through the cuticle of the stigma. Pollen hydration and ocreatine formation may be inhibited when the pollen is incompatible. The pollen tubes of a species show strong phenotypic differences during their growth through the styles of different species. The apical end shows cytological anomalies when there are specific incompatibilities in the stigma, style, or embryo sac.

Fertilisation in Angiosperms

Fertilisation in angiosperms is a two-part process: one sperm fuses with the egg cell, giving rise to the zygote, a process called syngamy by Strasburger and others; another fuses with the two polar nuclei of the central cell, creating the endosperm (triple fusion). Usually the pollen tube enters the ovule through the micropyle (porogamy), through the chalaza (chalazogamy), or through the integuments (pleurogamy). The sperm fertilising the egg cell first penetrates through the filiform apparatus. The sperm is then transferred to a synergid cell and later to the egg cell. The fusion between the sperm and the egg cell plasmalemma results in a bridge through which the sperm enters the egg cell.

The fusion of one sperm nucleus with the egg produces a diploid zygote; the fusion of another sperm nucleus with the nuclei of the central cell produces the triploid endosperm, typical in 60 to 70 percent of the dicotyledons (monosporic embryo sac of the *Polygonum* type). In the other embryo sac types, the endosperm varies from 2n (*Oenothera* type) to 3n in the *Allium* and *Adoxa* types. The remaining embryo sac types develop a polyploid endosperm. The double fertilisation transforms the ovule into the seed.

Fertilisation can occur within the same flower (autogamy), between two flowers on the same tree (geitonogamy), or between flowers of different individuals (allogamy or xenogamy). Some species can have autogamy and allogamy on the same tree; this phenomenon is called alautogamy.

Reproductive System in Gymnosperms

The gymnosperm, which means naked seed, is a group of superior vascular plants in which the ovules and seeds are not enclosed within a carpel. They are exposed on *scales* or similar structures, equivalent to sporophylls. Among the gymnosperms, the Coniferales are the dominant group. Coniferales are a primarily perennial, woody species, many of which have economic value. In the Neotropics, species of frequent use and economic value are numerous, including *Pinus oocarpa* Schiede ex Schltdl., *P. montezumae* Lamb., *P. caribaea* Morelet, *P. maximinoi* H.E. Moore, *P. patula* Schiede & Schltdl. & Cham., *P. ayacahuite* C. Ehrenb. ex Schltdl., *Abies guatemalensis* Rehder, *A. religiosa* (Kunth) Schltr. & Cham. (Pinaceae), and the ciprecillos (*Podocarpus* guatemalensis, P. costaricensis, P. macrostachyus, Prumnopitys *standleyi* de Laub., Podocarpaceae).

Coniferales are predominantly monoecious and have ovules and pollen grains clustered in strobiles or cones of different shapes and sizes. All the species of Pinaceae are monoecious, but some families, such as Cupressaceae and Podocarpaceae, have both dioecious and monoecious species. The strobile has a central axis with numerous imbricate sporophylls (2n), called scales or bracts, distributed in a closed spiral. The male strobile (androstrobile) has microsporophylls; usually each microsporophyll has two pollen sacs (microsporangia) in the abaxial (lower) surface. However, *Cycas media* may produce more than 1,000 sacs, *Zamia floridana* several dozen, the Taxodiaceae 2 to 9, the Cupressaceae 3 to 6 (sometimes more), the Taxaceae 2 to 8, and the Araucariaceae 5 to 20.

In families such as Pinaceae, the female strobile (gynostrobile) has megasporophylls and each has two ovules inverted in the adaxial (upper) surface. Other families have more or fewer ovules per megasporophyll; for example, the Taxodiaceae have two to nine and the Cupressaceae two to many, while Araucariaceae, Podocarpaceae, and Taxaceae have a single ovule. The numerous male strobiles have a short life span; when mature they may have a yellowish, purplish, or reddish coloration. The less numerous female strobiles also show variations in color. The color is genetically determined and can vary with elevation; for example, the white fir (*Abies concolor*) has light green or dark purple cones. The morph with green cones is more common in low elevations, while the morph with dark purple cones becomes more numerous with increases in elevation.

Pollen

The microsporangium (2n) or pollen sac has wall, tapetum (in many species), and sporogenous cells. The latter give rise to the microspore mother cells or microsporocytes, each producing four microspores (n) by meiosis. The microspores remain inside the microsporocyte wall for variable periods of time depending on the species. During this period, the microspore divides three times and forms the microgametophyte or partially developed pollen grain.

The latter has two prothallic cells (they die early), a generative cell, and a tube cell. Pollen sac opening and pollen grain release usually occur in this stage. After pollination, the pollen grain germinates and develops the pollen tube, which penetrates through the nucel-lus to the archegonium, absorbing nutrients from nucellar origin. Usually, the generative cell divides during pollen tube penetration, giving rise to two or sometimes more sperm. In the cycads and *Ginkgo*, the sperm are multiflagellate; the conifers lack flagella and are steady.

In genera such as *Pinus*, *Podocarpus guatemalensis*, and *Prumnopitys standleyi*, the pollen has air vesicles or wings; in others, such as *Pseudotsuga*, the pollen grains are spherical or oval with a smooth wall and lack wings. The pollen is yellow and abundant.

Ovule

The ovule has one integument fused to a multicellular body called the nucellus (functionally equivalent to the megasporangium). In the apex (distal end) the integument forms the micropyle. The proximal end, opposite the micropyle, is the chalaza. In the micropylar end of the megasporangium, a megasporocyte or megaspore mother cell (2n) develops. Through meiosis, this cell originates a linear tetrad of megaspores (n); only the innermost megaspore is functional. Inside the functional megaspore, numerous free nuclear divisions occur (e.g., up to 2,000 nuclei in the Pinaceae); cell wall formation occurs later. It begins in the periphery and proceeds centripetally. A cellular megagametophyte forms and is surrounded by the megaspore wall, which frequently increases in thickness. The megagametophyte cells are rich in nutrients. During or after cell wall deposition, some surface cells in the megagametophyte, usually placed at the micropylar end, form varying numbers of archegonia (e.g., 2 to 6 in *Pinus*, 2 in *Ginkgo*, up to 60 in *Sequoia*). A jacket of sterile cells enclosing and protecting an egg cell placed at the canal base forms each archegonium.

Pollination and Fertilisation

Pollination is the transport of the partially developed endosporic microgametophyte (pollen grain) from the pollen sac to the micropyle. The developmental stage in which polli-nation occurs in gymnosperms varies. In most conifers, taxads, and *Ginkgo* it occurs when the ovule has only sporogenous cells or megaspore mother cells. During the prepollination phase the micropyle is open and many cell divisions occur in the ovule. In many species, a pollen chamber is formed at the micropilar end through degeneration of nucellar cells.

The wind is the main pollination vector, although several species are entomophilous. Coleopterans and hymenopterans are the most frequent agents. Examples of entomophily include *Zamia pumila, Macrozamia communis*, and *M. riedlei*. The angiostrobiles of some cycads release a strong odor, which helps to attract the insects. Many conifers, such as *Pinus*, have pollen adhering to the integument and floating later in the liquid exuded by the ovule (pollination drop). While the drop evaporates, the pollen is mobilised to the base of the micropylar canal where it finds the egg cell. In other species, such as *Larix occidentalis* and *Pseudotsuga menziesii* (Mirb.)

Franco, the pollen is captured and detained by a stigmatic extension with many trichomes. The trichomes produce a small secretion that contributes to the initial development of the pollen tube. The pollen tube enters the megaspore wall and elongates to reach the archegonium where it establishes. Usually, a long period of time elapses between pollination and fertilisation. In this period the ovule undergoes several changes: the closing of the micropylar canal and the increasing in size, through cell division and activation, of the spongy tissue or tapetum. The latter originates from sporogenous cells not functioning as spore mother cells or from nuclear cells surrounding the sporogenous tissue. Later, this tissue degenerates and collapses.

The primary difference between gymnosperms and angiosperms is that fertilisation in the former is not a twopart process. A single fertilisation occurs in which one sperm fuses with the egg cell and the zygote (2n) develops. The remaining sperms degenerate and no true endosperm is formed. Usually, the egg cells of several archegonia are fertilised; however, only a single embryo develops. The reserve materials stored in the female gametophyte nourish the embryo. The ovule is usually the same size as the mature seed. *Ginkgo* and the cycads have an haustorial pollen tube. Double fertilisation has reportedly occurred in several gymnosperms such as *Ephedra, Gnetum, Pseudotsuga, Thuja,* and *Welwitschia*; nevertheless, double fertilisation has been corroborated only in *Ephedra nevadensis*.

The Fruit

Angiosperm Fruit

The fruit is the structure containing the seed. It develops from the gynoecium of the flower, which is frequently associated with other floral organs. The process of fruit development has four stages:

(1) Initiation and development of the floral bud leading to the formation of the mature flower, with one or several ovules in the gynoecium.

(2) Cessation of cellular division and elongation during pollination, pollen tube development, and ovule fertilisation. After pollination, many flowers close the corolla or undergo a fast collapse of the corolla and the other floral organs. Floral collapse is correlated to increasing respiration and ethylene production.

(3) Postfertilisation events: Growth of the ovary wall and associated tissues, by cell elongation and ovary wall transformation into the fruit wall or pericarp.

    Seed development from the fertilised ovule; however, some fruits may develop without fertilisation or seed development (parthenocarpy). In general, seed development and perianth and stamen withering and abscission are simultaneous.

(4) Fruit ripening followed by senescence and, sometimes, dehiscence and abscission.

The pericarp can be more or less differentiated and frequently shows two or more different layers. If the layers are distinguishable, they are called (from the outside to the inside): exocarp (epicarp), mesocarp, and endocarp. These terms are used for descriptive purposes and do not relate to the ontogenetic origin of the layers.

Numerous fruits have sigmoid growth patterns, which start with an exponential increase in size, diminishing later to adjust to a sigmoid pattern. Other fruits have more complex growth patterns involving periods of growth with an interval of reduction or detention of growth. The fruits of many tropical trees have the latter type, as well as some crops with edible drupes.

Fruit growth requires cell division and cell elongation. The range of cell division differs from one fruit to another. Some complete division during pollination while others extend it into the postpollination period. The reserves stored by fruits are synthesised in the leaves and transported by the phloem. The existence of haploid, diploid, and triploid tissues (sporophyte 2n, embryo 2n, but differing from the sporophyte genome; the endosperm usually 3n, and the perisperm 2n) complicates the growth pattern. During fruit development, the differential growth follows different directions in the distinct parts and tissues of the fruit.

Normally, fruit development depends on and is controlled by growth regulators synthesised in the developing seeds. The seeds produce auxins, gibberellins, cytokinins, abscissic acid, and ethylene. The relationships between these substances are very complex but the most important fact is the balance of regulators at the different stages of development. In many species a fruit lacking seeds can develop well; for example, the empty and filled fruits (samaras) of *Terminalia amazonia* and *T. oblonga* Steud. are indistinguishable externally. Generally, the extraction of fertilised ovules from a developing fruit stops development.

Fruit shape usually reflects the internal distribution of the seeds (e.g., *Cojoba arborea*), and seed shape frequently reflects the number of seeds inside the fruit (e.g., *Carapa guianensis, Eschweilera panamensis, E. costaricensis*). Seed number inside the fruit and final fruit and seed size are correlated. In many cases, the fruit shape reflects the internal pressures exerted by the overgrown seed(s) developing inside it; in this case, embryo growth is limited by the pericarp (e.g., *Prioria copaifera* Griseb., *Dipteryx panamensis*). Fruit maturation is the ensemble of processes associated with the attainment of maximal size and the qualitative transformation of the tissues. The latter involves tissue softening as a product of hydrolytic conversions of stored reserves, as well as changes in pigmentation, production of tastes, and disappearance of astringent substances (e.g., *Achras, Elaeoluma, Manilkara, Micropholis, Pouteria,* Sapotaceae).

The hydrolytic changes in the reserve materials result in the production of sugars by starch and lipid chemical transformation. Proteins follow an opposite tendency; their synthesis intensifies during the process of ripening. Enzymes present in the tissues soften the fruit tissues and the process requires that pectic substances present in the cell walls become soluble. Fruit ripening promotes drastic changes in the respiratory rate. The increase in respiration known as climacteric respiration is related to ethylene concentration. Other events associated with fruit ripening include the increase in ribonucleic acid (RNA) and the change in cellular permeability. Climatery is the developmental period of certain fruits (especially fleshy fruits) in which many biochemical changes occur that determine the transition from growth to senescence and ripening. These changes are promoted by the autocatalytic production of ethylene.

Fruit Abscission

Abscission is the organised separation of cells resulting in tissue separations; this occurs in leaves, flowers, fruits, and stems. Abscission occurs during the opening of many types of fruits (fruit dehiscence). The actions of growth regulators cause abscission at different stages of fruit development. When it occurs at maturity, the fruits may contain seeds (indehiscent fruits; e.g., *Minquartia guianensis, Prioria copaifera*). The fruit usually lacks seeds if the dehiscence precedes abscission (e.g. *Cedrela*).

The abscission zone varies in the different fruits and a fruit may have more than one zone. For example, apples abscise at the base of the pedicel, while *Prunus* (prunes), *Calophyllum brasiliense*, and *Minquartia guianensis* first separate at the fruit base and then at the pedicel base. The degree of cell differentiation in the abscission zone also varies from one fruit to another. Abscission is an active process involving the separation of the middle layers of the cell walls in the abscission line, under enzymatic action. Usually a cellulase and a polygalacturonase are involved. The abscission process requires the synthesis of proteins; e.g., the ethylene stimulates the abscission and induces the synthesis of enzymes in the cell wall.

Methods of Dehiscence

The spontaneous breaking of the pericarp allowing seed dispersal is known as dehiscence. Methods of dehiscence are diverse. If the ovary derives from a single carpel, the pericarp may break longitudinally along the ventral suture (suture joining the carpel margins), along the dorsal suture (midvein line), or along both sutures. In ovaries with two or more carpels, tissue separation takes place along the lines (septa) joining contiguous carpels; this type of dehiscence is called septicidal. Rupture along the carpel backs (middle of the loculus) where the dorsal bundles are located is loculicidal dehiscence. In some fruits the dehiscence that occurs along a horizontal ring involving all carpels is known as circumscissile dehiscence. Dehiscence can also take place along one or several pores (poricide dehiscence) which sometimes have an operculum (poricidal-operculate dehiscence).

Classification Fruits

Flower organisation in terms of carpel number, distribution, degree of fusion, and structure influences fruit size, shape, texture, and anatomy. Fruit classifications are artificial and emphasise certain anatomical and biological characteristics, such as fruit texture (dry or fleshy) and dehiscence (dehiscent and indehiscent). Although these characteristics are valid for classification purposes they lead to juxtapositions and duplications of types in some classification systems. The fruits are separated into three categories: sim-ple, aggregate, and multiple. A simple fruit is originated by a single gynoecium, unicarpellar or syncarpic. The aggregate fruit is derived from an apocarpic gynoecium; each carpel keeps its identity until ripening. The multiple fruit derives from an inflorescence; that is, from a combination of the gynoecia of many flowers, which are sometimes coalescent. If any of these fruits has extracarpellar tissue it is an accessory fruit.

Simple Fruits

Pericarp texture is the basis used to categorise the simple fruits into dry (dehiscent and indehiscent) and fleshy fruits.

Dry Indehiscent Fruits

Those fruits that remain closed at maturity. They usually arise from ovaries in which only a single seed develops, although in some cases more seeds are present. This category includes several types of fruit:

Achene: Small fruit bearing one seed fused to the fruit wall at a single point; it derives from a superior ovary with only one locule; e.g., the sunflower *(Helianthus annuus,* Asteraceae).

Balausta: Fruit with numerous locules and numerous seeds; the pericarp is coriaceous; e.g., *Punica granatum* (Punicaceae).

Calybium (balano): Hard fruit with only one locule developed in the mature fruit. The calybium derives from an inferior ovary; e.g., the oaks *(Quercus, Fagaceae).* The oaks have tricarpellar, trilocular flowers with two ovules per locule, but only one locule and one seed develop. The *Quercus* fruits have a dry, cupuliform involucre; because of that the fruit is sometimes classified as an accessory fruit under the name of cupule or glans.

Indehiscent capsule: Fruit derived from an ovary with two or more locules; e.g., *Crescentia cujete,* Amphitecna sessilifolia (Bignoniaceae); Catostema fragrans (Bombacaceae); Apeiba tibourbou, Heliocarpus appendiculatus (Tiliaceae); and *Simira maxonii* (Rubiaceae).

Nut: Fruit with only one seed. The pericarp is hard. Usually the nut derives from a unilocular ovary; e.g., *Anacardium excelsum* and *A. occidentale* (Anacardiaceae). In A. occidentale the fleshy structure corresponds to the pedicel.

Samara: Winged fruit or diaspore; e.g., Anacardiaceae (Astronium graveolens), Bombacaceae *(Cavanillesia platanifolia),* Combretaceae *(Terminalia amazonia, T. oblonga),* Oleaceae *(Fraxinus uhdei* (Wenz.) Lingelsh.), Fabaceae-Caesalpinioideae *(Sclerolobium costaricense, Tachigali versicolor),* Fabaceae-Papilionoideae *(Dalbergia retusa, Hymenolobium mesoamericanum, Platymiscium pinnatum* (Jacq.)

Dugand, P. pleiostachyum, Myrospermum frutescens, Myroxylon balsamum (L.) *Harms, Paramachaerium gruberi, Pterocarpus* hayesii), *Polygonaceae* (Triplaris surinamensis), and *Ulmaceae*. Some authors call the samara of different Fabaceae samaroid pods.

Camara: Unicarpellar fruit, indehiscent or late dehiscent. The fruit pericarp and the seedcoat are independent. Frequent in legumes and commonly described as an indehiscent pod; e.g., *Fabaceae-Caesalpinioideae, Fabaceae-Mimosoideae, Fabaceae-Papilionoideae.*

Dry Dehiscent Fruits

These are fruits that open at maturity. The following types are found in this category:

Capsule: Fruit derived from an ovary with two or more carpels. The pericarp opens along predetermined lines as the result of tensions produced by tissue drying. Several kinds of capsules exist:

Acrocidal: The capsule opens through fissures or specific openings; e.g., *Gossypium hirsutum.*

Circumscissile (pyxidium): The cover is named operculum and the dehiscence line is a well-defined circumference.

It is common in Lecythidaceae; e.g., Lecythis ampla, Eschweilera panamensis, E. costaricensis, and Couratari guianensis Aubl.

Loculicidal: The capsule opens longitudinally in the locule cavity (carpel midvein region); e.g., Tiliaceae (Luehea seemannii), Bixaceae (Bixa orellana), Bignoniaceae (*Jacaranda copaia* (Aubl.) D. Don, *Tabebuia rosea* (Bertol.) DC., *T. chrysantha*), Elaeocarpaceae (Sloanea faginea, S. latifolia), Loganiaceae *(Placospermum buxifolium),* Meliaceae *(Guarea, Trichilia),* Sapindaceae *(Cupania),* and Turneraceae *(Erblichia odorata).*

Poricidal: The capsule opens through a pore. If the pore has an operculum the capsule has poricidal-operculate dehiscence; e.g., *Triodanis.*

Septicidal: The capsule opens longitudinally through septa. It is common in Meliaceae (Cedrela, Carapa guianensis, Swietenia), and Myristicaceae (Virola, Otoba, Myristica fragrans).

Follicle: Fruit derived from a carpel opening along a suture; e.g., Proteaceae *(Roupala montana, Macadamia integrifolia).*

Legume or pod: Coriaceous fruit derived from a monocarpellar gynoecium. It opens along the ventral and dorsal sutures. Typical fruit in the *Fabaceae-Caesalpinioideae (Copaifera aromatica* Dwyer, *Peltogyne purpurea),* and Fabaceae-Mimosoideae *(Cojoba arborea, Abarema macradenia).* The dehiscence is explosive in *Pentaclethra macroloba* (Fabaceae-Mimosoideae) and *Gliricidia sepium* (Fabaceae-Papilionoideae).

Loment: Pod or legume splitting into segments each containing a seed; e.g., *Muellera frutescens* (Fabaceae-Papilionoideae) and *Senna skinneri* (Fabaceae-Caesalpinioideae). The last species is extremely specialised; the pod has deep constrictions in the break sites.

Schizocarp: Fruit splitting into units or fruitlets containing the seeds; e.g., *Hevea brasiliensis* Müll Arg. and *Hura crepitans* (Euphorbiaceae). The mericarp (cremocarp, carpopodium) are the units; if the unit is equivalent to half a carpel it is called a carcerule.

Fleshy Fruits

Those fruits derived from a monocarpellar or multicarpellar gynoecium. The pericarp has carpellar or extracarpellar tissue (hypanthium, perianth, androecium, peduncle, bracts). The pericarp can be fleshy in the external or internal parts; in some fruits both zones are fleshy. In some fruits the placenta and the partitions of multilocular ovaries are also fleshy. The principal fleshy fruits follow:

Pome: Fruit derived from a gynoecium with two or more carpels and an inferior ovary. The receptacle and hypanthium are adnate to the ovary and form most of the fleshy tissues.

The endocarp is cartilaginous; e.g., Rosaceae (*Pyrus* malus, Pyrus communis, Cydonia oblonga).

Berry: Fruit with a succulent pericarp, e.g., Myrtaceae (Psidium friederichsthalianum, P. guajava, Syzygium jambos, Myrcianthes fragrans), Clusiaceae (Calophyllum brasiliense, Symphonia globulifera), Lauraceae (Ocotea, Nectandra, Persea), and Rubiaceae (Posoqueria).

8

Technology Involvement in Modern Seed Production

Biotechnology is providing powerful and useful tools, in a continuum of technical evolution that contributes or could contribute to the improvement of crop production, food quality and safety, while preserving the environment.

Biotechnology in Planting Material Propagation

Biotechnological tools have greatly contributed to the production and supply of improved quality seed and planting material to farmers worldwide. Among other uses, biotechnology is employed to:

- speed-up the multiplication process for vegetatively propagated crops,
- detect diseases transmitted by seed or planting material,
- eradicate diseases transmitted by planting material,
- protect seed with biological control agents, and
- test varietal identity and purity.

Tissue Culture

Miniaturised *in vitro* multiplication of plant material under aseptic and controlled artificial conditions, also known as micropropagation, has been used for decades to speed up the propagation process for several vegetatively propagated crops. This is the case for fruit trees (e.g. banana, date palm), roots and tubers (e.g. potato, cassava), vegetables (e.g. strawberry, asparagus), and ornamentals (e.g. roses, orchids). Many companies and institutions worldwide have invested or specialised in this activity, to provide farmers and growers with high quality and healthy planting material.

Somatic embryogenesis, a variation of micropropagation (where embryos are directly regenerated instead of shoots and roots), is being used widely for oil palm. In the early 1980s, several organisations investigated the possibility with some crops to use somatic embryos that could be encapsulated with different chemical and biological compounds. These "artificial seeds" would have been planted and treated as seeds. Potential benefits would have been tremendous from a plant breeding, seed production and seed treatment point of view. However, technical and economic constraints have not allowed commercial development of this technology.

Detection of Diseases Transmission

Two methods are being more and more used for detection of diseases transmitted by seed and planting material:

— ELISA (Enzyme Linked Immunosorbent Assay) is a widely used technique based on an antibody-antigen reaction, and used for diagnostic determination of the presence of specific molecules in a sample.

— PCR (Polymerase Chain Reaction) is a technical procedure that amplifies and makes it possible to detect a specific piece of DNA.

These two techniques are much more sensitive and reliable than conventional seed health assays on grow-out media, and will ensure a much higher seed health level. However, for the time being, they are only operational for a limited number of diseases transmitted by seed or planting material. Moreover, the high sensitivity of these techniques, in particular PCR techniques, may lead to false-positive results. As far as quarantine pests are concerned, misuse of these techniques may lead to unnecessary troubles in international seed trade. To limit these potential difficulties, and optimise potential benefits from these new technologies, ELISA and PCR techniques must be used with care and have to be associated with appropriate sample sizes and thresholds, to be set jointly by scientists and quarantine authorities.

Eradication of Diseases Transmission

A major problem of vegetatively propagated crops is transmission of diseases, in particular of viruses, through planting material. Several cultivars of garlic, potato, apple tree, etc., have been known to be totally contaminated by viruses. In the absence of genetic resistance within the genepool, meristem culture has been widely used to transform these cultivars into healthy cultivars, free from diseases. The principle is that a meristem (the apical part of a stem) is normally free from diseases. If a meristem is taken from a contaminated plant and is grown *in vitro* under appropriate conditions, it will regenerate a plantlet that will be disease-free. This plant can then be propagated either *in vitro*, or *in vivo* under strictly controlled conditions, to ensure production of healthy planting material. This technique has been widely used for decades on many vegetatively propagated crops and is still used to regenerate some contaminated cultivars. Its association with disease detection methods such as ELISA or PCR techniques makes it a very powerful tool to ensure propagation of healthy planting material.

Use of Biological Control Agents

Biological control agents used as seed treatments are micro-organisms that protect seeds and seedlings from various pathogens. Of the biological control agents patented by early 1999, 84% were bacteria (mainly *Pseudomonas* spp. and *Bacillus* spp.) and 16% were fungi. The mode of action of biological control agents can be categorised as antagonism, antibiosis, competition or mycoparasitism. These agents may provide a good solution for protection against specific pathogens, in particular in the case of organic farming. Cotton has been the first large-scale agronomic crop treated with biological control agents for the suppression of seedling diseases of the rhizosphere, and, today, much of the cotton planted in the USA is treated with such agents. These products have also been used and tested on several other agricultural and vegetable crops.

Limited understanding of rhizosphere ecology, formulation difficulties, storage and stability problems after application to the seed, and release of some products that sometimes did not meet performance expectations, have slowed down the growth and adoption of the technology. Moreover, regulatory issues, such as registration and discard issues, may constitute constraints to the development and use of such products. Today, they are being actively developed by a number of companies. If the investment in this technology continues at the current pace, these products will probably have a significant place in the options to protect seed and seedlings.

Control of Variety Identity and Purity

Cultivars are described using phenotypic characteristics, i.e. using characteristics resulting from the expression of genes. These characteristics may be either morphological or physiological. They may also be electrophoretic, as in the case of isozymes. Using isozymes to check varietal identity and purity provides a great advantage to seed producers, since it is a cheap and simple technique, and it is possible to have test results without planting *a posteriori* control plots, contrary to the analysis of other phenotypic characteristics. Moreover, if performed under standardised conditions, isozyme electrophoresis results are unaffected by genotype environment interactions. Nevertheless, to make it possible to distinguish varieties with accuracy, this technique requires several markers, isozyme polymorphism, and cultivars that are sufficiently uniform for these characteristics (unless they were bred for). Today, this technique is used for some crops, such as maize.

Biotechnology in Plant Breeding

Crop improvement is the exploitation of genetic variability, followed by several generations of selection. Breeders have always used the most modern technologies available to them. This has permitted them to make considerable progress during the last twenty years, thanks in particular to the development of biotechnology. These tools permit:

— an acceleration of the selection process,

— new genetic combinations that are not possible through conventional breeding, and

— greater precision in the desired modifications of the genome.

Doubled Haploids

Using *in vitro* techniques, it is possible to regenerate plants from pollen or ovules. These plants, which contain only one copy of each chromosome, are called haploids. They are not viable. After appropriate chemical treatment, it is possible to restore the normal number of chromosomes and to regenerate viable plants. These plants, called double-haploids, are homozygous for all their genes. Such plants are of tremendous interest to plant breeders, since they allow development of pure line varieties or inbred parental lines much more quickly than through conventional breeding.

Androgenesis (regeneration from pollen) has been successfully used for crops such as eggplant, pepper and wheat. Gynogenesis (regeneration from ovules) is used on barley. However, the bulbosum method used with barley does not require *in vitro* cultivation of ovules; development of haploids is obtained *in vivo* through interspecific crosses between barley and *Hordeum bulbosum*, a wild relative.

Marker-assisted Breeding

Markers may be either phenotypic or genotypic, and marker-assisted breeding developed in the 1980s with the evolution of DNA marker technologies. Today, the main DNA markers used in breeding programmes are Random Amplified Polymorphic DNA (RAPD), Amplified Fragment Length Polymorphism (AFLP), microsatellites, and Expressed Sequence Tags (ESTs). Each of these markers has a different set of advantages and limits. Cost and possible automation of the techniques are of particular importance for their adoption.

Use of molecular markers, in association with linkage maps and genomics, offers plant breeders the potential to make genetic progress much more precisely and rapidly than through phenotypic selection. It also offers the possibility of addressing previously unattainable goals.

There are many applications for the use of DNA markers in breeding programmes, which fall into four broad groups, based on the purpose of the intervention:

— enhancing knowledge of breeding material and systems, such as better understanding and more effective breeding of Quantitative Trait Loci (QTL);

— rapid introgression or backcross breeding of simple characters, as the number of back-crosses required can be reduced drastically if there are markers for the character to be introduced and for the genetic background of the recurrent parent;

— early or easy indirect character selection, which is important for genes that cannot be detected at an early development stage, such as high lysine and tryptophan genes in maize; and

— new goals not possible through traditional breeding, including pyramiding of disease resistance genes with indistinguishable phenotypes.

Embryo Rescue

Breeders need access to the largest possible genetic variability. In some cases, variability available within a given species is not sufficient to answer a specific problem (e.g. resistance to some new disease). A solution available to breeders is inter-specific hybridisation (crossing plants from separate but related species). However, embryos resulting from such hybridisation rarely survive, due to incompatibilities between the embryo and the mother plant. Saving embryos is sometimes possible through their *in vitro* cultivation, which make it possible to isolate the inter-specific embryo from the hostile mother plant environment.

This technique has been used for the introduction of disease resistance into squash, lettuce, tomato, etc. This technique will be replaced by transgenesis, which will provide the same result, but much faster and with much more accuracy. Indeed, contrary to inter-specific hybridisation, transgenesis leads to the introduction of solely the target gene, and eliminates the need for several generations of backcrossing.

Protoplast Fusion

Fusion of protoplasts is another technique to allow inter-specific hybridisation between species that cannot be crossed through conventional breeding, even using *in vitro* embryo rescue. Protoplasts are plant cells that have had their outer walls removed through chemical treatment. While it is difficult or impossible to fuse plant cells, it is possible through various techniques (using either chemical or physical treatments) to merge protoplasts from different crop species or genera, and then to regenerate a whole plant resulting from the fusion process.

This technique has been used to introduce traits such as male sterility into rapeseed, or disease resistances in potato. Similarly to embryo rescue, this technique will most probably be replaced by transgenesis, which is a faster, more effective and more precise technique. Moreover, protoplast fusion is generally not effective beyond the family level due to incompatibilities between two too-distant genomes making it impossible to regenerate plants.

Transgenesis

Transgenesis (also called genetic transformation or genetic engineering) is the most recent step to increase genetic variability available within a crop. Transgenesis aims to introduce, through different techniques, a specific gene (included in a gene construct) from a donor species into the genome of a host species. Organisms resulting from transgenesis are commonly called Genetically Modified Organisms (GMOs).

Transgenesis is being used to introduce a broad range of new agronomic, processing and nutritional traits into the main agricultural and vegetable crops. In comparison to embryo rescue and protoplast fusion, transgenesis is not constrained by reproduction barriers. Genes can be transferred from one realm into another. Moreover, only the specific gene construct is introduced in the host organism. This provides great precision and rapidity to the enhancement process. Transgenesis is a very promising tool for the development of new varieties with specific traits that are not present within the crop genepool.

Genetically Modified Organisms

The first GM crop approved for cultivation was the "Flavr Savr" tomato in 1994 in the USA. Since then, development of transgenic crops has been incredibly fast (Table 1). In 1999, the area cultivated with transgenic crops reached 39.9 million ha. The total market for transgenic seed was assessed at approximately US$ 3 000 million, out of a global commercial seed market amounting to US$ 30 000 million.

Table 1. Evolution in the use of transgenic crops

Year	*Acreage (million ha)*	*Market Value (US$ million)*
1995	-	1
1996	1.7	152
1997	11.0	851
1998	27.8	1,959
1999	39.9	~ 3000

In 1999, transgenic cultivars were commercially released in at least 12 countries (Table 2), with trials ongoing in several others. In 1999, of a global total of 39.9 million ha of transgenic crops planted, USA, Argentina and Canada cultivated 28.7, 6.7 and 4.0 million ha, respectively. In eastern Europe, Romania and Ukraine grew transgenic crops for the first time in 1999.

Table 2. Cultivation of transgenic crops in 1999, by country

Country	*Acreage (million ha)*
USA	28.7
Argentina	6.7
Canada	4.0
China	0.3
Australia	0.1
South Africa	0.1
Mexico	< 0.1
Spain	< 0.1
France	< 0.1
Portugal	< 0.1
Romania	< 0.1
Ukraine	< 0.1

In 1999, the main GM crops were soybean, maize, cotton and rapeseed, with 21.6, 11.1, 3.7 and 3.5 million ha, respectively (Table 3). Transgenic soybean represented 30% of the global soybean acreage. This rate was 14%, 10% and 8% respectively for rapeseed, cotton and maize. Transgenic cultivars of potato, squash and papaya were also commercialised in 1999, and trials were underway on numerous other crops.

Table 3. Dominant transgenic crops in 1999

Crop	*Trait*	*Acreage (million ha)*	*Transgenic area as proportion of global area (million ha)*
Soybean	Herbicide tolerance	21.6	30%
Maize	Insect resistance (*Bt*)	7.5	8%
	Bt + Herbicide tolerance	2.1	
	Herbicide tolerance	1.5	
Cotton	Herbicide tolerance	1.6	10%
	Insect resistance (*Bt*)	1.3	
	Bt + Herbicide tolerance	0.8	
Rapeseed	Herbicide tolerance	3.5	14%

The main commercially released traits in 1999 were of agronomic interest, the so-called "input" traits. Herbicide-tolerant crops were grown on 31.1 million ha. and insect-resistant crops on 11.7 million ha, of which 2.9 million ha were both herbicide tolerant and insect resistant.

Input traits that are in the pipeline for release in the next few years include fungal, bacterial and virus resistances, delayed senescence, and hybridisation genes. Other crops have been transformed and are being commercialised on limited acreages, or expected to be commercially released soon, including beet, rice, wheat, sunflower, sugar cane, tomato, pepper, sweet potato, cassava and banana.

In the short to medium term, transgenesis should be increasingly used to transform processing characteristics of crops, the so-called "output" traits. The main traits will be:

— to remedy to deficiencies, including higher vitamin content in soybean, rapeseed and rice, or higher iron content (in rice);
— improved nutritional profile, such as improved amino acid composition (maize), or improved fatty acid composition (maize, soybean, rapeseed);
— improved processing properties, including modified starch (maize, potato), higher solid content (potato), or improved fibre quality (cotton); and
— reduction of post-harvest losses, such as delayed ripening (papaya), or improved storage capacity (potato).

Benefits and Risks of GMOS

Benefits and risks of GMOs are assessed through comparison of the new organism with its "conventional" counterpart (variety or food) and associated techniques (e.g. pest management or food processing techniques).

Input traits, in general, allow lower use of pesticides and, as a consequence, they benefit the environment and improve farmers' revenue. Among other specific benefits, insect-resistant varieties limit post-harvest losses (insects are the origin of up to 50% of loss of the harvested product in developing countries) and production of mycotoxins (responsible for serious health problems), and herbicide-tolerant varieties contribute to limit soil erosion. Output traits will be of considerable benefit to consumers through access to healthier food. Therefore, farmers, the environment and consumers benefit from the development of GMOs. This benefit will increase with the release of GMOs combining input and output traits.

Today, despite the fact that approximately 80 million ha have been grown with GMO crops worldwide from 1995 to 1999, and that billions of people have eaten GMO foods without any documented harmful effect on human health or the environment, we have to be aware that GMOs, as any new product, may be associated with some risks, which may be of various kinds, e.g. use of herbicide-tolerance genes could result in production of "super weeds" that would be resistant to total herbicides; introduction of a gene from a species that is known to be allergenic could result in the introduction of allergens in the novel food. Moreover, some aspects, which are not specific to GMOs, have also to be taken into account, such as the risk of insects developing resistance to *Bt* genes. These risks are taken seriously into consideration during the pre-release risk analysis and, where needed, specific risk management procedures may be established to minimise these risks.

Regulatory context

Several intergovernmental organisations have adopted or are negotiating international instruments addressing the biotechnology issue, and more specifically the GMO issue. Those most relevant to the seed sector are:

— The Conference of the Parties to the Convention on Biological Diversity (CBD) adopted in 2000 the Cartagena Protocol on Biosafety, aimed at protecting biological diversity from possible adverse effects of Living Modified Organisms

(LMOs) moving across borders. This Protocol is not likely to enter into force before 2002.

— Codex Alimentarius has established a Task Force on food derived from modern biotechnology, which met for the first time in March 2000, to negotiate guiding principles on food safety analysis.

— The Interim Commission on Phytosanitary Measures (ICPM), a body established under the revised International Plant Protection Convention (IPPC), convened a meeting in 2000 to discuss what should be the role of plant protection organisations vis-à-vis the implementation of the Cartagena Biosafety Protocol. It could result in specific guidelines.

— The Commission on Genetic Resources for Food and Agriculture (CGRFA) initiated in the early 1990s negotiations on a Code of Conduct for Plant Biotechnology, aimed at maximising positive effects and minimising possible negative effects of biotechnology. This work was interrupted, but was expected to restart in 2001.

— The Organisation for Economic Co-operation and Development (OECD), through the OECD Schemes for the Varietal Certification of Seed Moving in International Trade, initiated in 1999 discussions on an experimentation and regulatory framework regarding the issue of adventitious presence of GMOs in non-GM seeds.

Risk analysis framework

Before release, any new GMO has to go through a thorough environment and food safety analysis, in addition to variety registration procedures (where needed). Risk analysis has three components: risk assessment; risk management; and risk communication. Risk analysis is based on sound scientific procedures. It shall consider, among others, the recipient and donor organisms, the vector, the inserted gene, the gene construct, the resulting GMO, its intended use, the receiving environment, etc. When needed, specific management procedures may be required. In case of potential risks that cannot be prevented or mitigated to an acceptable level, through appropriate risk management procedures, the GMO shall not be released. Risk analysis is, in general, the result of a joint effort of the GMO developer (private or public) and of public regulatory agencies.

Issues in GMOs development

Despite careful scientific analysis before release of GMOs, which has proven its efficiency, several political issues surround GMO development:

Precautionary principle vs precautionary approach: Precaution, or the precautionary approach, is part of any research and development (R&D) activity. Before being released, any new product is subject to a scientifically sound risk analysis. Therefore, the precautionary approach is not a new concept. In contrast, the precautionary principle is a rather different concept, used by opponents to modern technologies, to argue that lack of scientific knowledge or scientific consensus should be interpreted as indicating a particular level of risk. This discussion, which may have an impact on development of modern biotechnology, although not specific to it, is a hot topic in various forums.

Traceability: European countries would like to trace GMOs from seed production to the consumer's plate, so that consumers can be informed regarding the food origin, and in order to be able to quickly react in case of any possible problem. This vision is not supported by all countries, in particular due to the cost implications of traceability. Developing countries have also expressed concerns regarding implementation of this concept in their specific socio-economic environment.

Substantial equivalence: Substantial equivalence is a concept used to identify similarities and differences between the GM food and a comparator with a history of safe use, and which subsequently guides the safety assessment process. This concept was developed due to the complexity of food composition, which required development of a specific concept (it was not possible to adopt models used to assess safety of single chemical compounds). Despite criticism of this concept, experts recognised that there were currently no alternative concept or strategy that would provide a better assurance of safety for GM foods.

Labelling: This issue is the subject of a very political debate, with two opposite points of view: on one hand, those who want labelling on the process and, on the other hand, those who want food products to be labelled only for their new characteristics, if any.

Antibiotic resistance marker genes: Such genes were used in the early stages of genetic engineering and therefore in the first GM varieties. Researchers are working on alternatives that will soon replace these marker genes. Nevertheless, experts agree that antibiotic resistance marker genes currently used in GM crops have been previously reviewed for safety and that there is no evidence that markers currently in use pose a health risk to humans or domestic animals.

Adventitious presence of GMOs in non-GM seeds: Depending on the crops, the seed market is either national, regional or global. Interdependence of countries for seed supply is particularly important for maize, herbage crops, vegetable crops and some industrial crops, such as rapeseed, cotton and soybean. As regards maize, much seed is moving across oceans. However, with respect to the important acreage of transgenic maize in some maize seed exporting countries, and to the fact that maize is a cross-pollinated species, it is more and more difficult to guarantee absence of GM seed in non-GM maize varieties. This incidental and involuntary presence of GM seed is called "adventitious" presence. In 2000, Switzerland was the only country that had adopted a threshold for adventitious presence of GMOs in both seeds and food products; some countries had only adopted a threshold for food products; and some had not yet adopted any threshold. Moreover, the detection technique is not yet standardised. A joint initiative of FIS, OECD and ISTA has launched an experiment on maize, cotton and rapeseed to develop an agreed appropriate sampling and detection protocol for a 1% presence threshold.

Unapproved events: Some transgenic events are approved in some countries, and not in others. Adventitious presence of these events not approved in an importing country, is not accepted. Tolerance for these events is zero (similarly to quarantine pests), but this requirement is impossible to implement by the industry. The problem is that events move with seeds and ignore geographical barriers. This issue is expected to create serious problems for the seed trade in the near future, and to probably lead to seed supply disruption for some crops in some countries (e.g. cotton seed in Greece, which is mainly imported).

Evolution of GM Plants

Today, it is very difficult to predict the evolution of GM plants acreage by 2010. It will be highly influenced by a combination of technical, regulatory, political and ethical factors. Development of output traits will probably have a positive impact on consumer acceptance in Europe. GMOs will probably develop rapidly in developing countries producing food for their domestic market (e.g. China, India), contrary to countries exporting commodities to Europe. Current negotiations to develop international binding regulatory instruments will also have an important impact, and it is difficult to predict the outcome of current and future negotiations. Indeed, the negotiators are under pressure from an active green lobby, which is anti-modern technology.

Technologies use in the Pipeline

Genomics

Genomics is the science aimed at identifying the entire set of genes of organisms. Progress in genomics is going very fast, thanks to developments in DNA sequencing techniques. As far as crop genomics are concerned, the most advanced crops are rice and maize. In April 2000, Monsanto announced that they had decoded the rice genome sequence to the level of a "working draft." This is the first crop genome to be described in such technical detail. This information will be shared with the International Rice Genome Sequencing Project (IRGSP), a ten-member consortium of rice genome sequencing projects worldwide. Pioneer Hi-Bred has also decoded over 80% of the maize genome. For dicotyledenous plants, research is less advanced. *Arabidopsis* has been used as a model species, and it was expected that its genome would be entirely decoded by the end of 2000.

Genetic Use Restriction Technologies (GURTs)

In nature, the expression of genes is regulated by several factors, which may be internal to the organism (e.g. proteins or other molecules resulting from the metabolism of the organism itself) or external (e.g. climatic factors). Modern biotechnology can also be used to regulate the expression of genes that are, for instance, not desirable at a certain stage of crop development. Methods that regulate gene expression are called Genetic Use Restriction Technologies (GURTs). GURTs are a specific domestication of the regulation of gene expression that occurs naturally in any organism.

Plant breeders have until now focused their activity on the introduction and recombination of genes. GURTs will allow them to work on the expression (or the non-expression) of genes at any given stage of crop development or in any generation.

Some potential applications of GURTs could be:

— increased production of specific molecules;

— regulation of the expression of resistance genes, so that resistance be expressed only when necessary;

— limitation of cross-pollination with other varieties, landraces or wild relatives; and

— improved protection of intellectual property rights.

GURTs are subject to an intense debate, in particular at the CBD level. The debate is highly emotional and focuses on impact of GURTs on biodiversity, food security and farmers' rights. If looked at scientifically and objectively, it is obvious that GURTs could provide solutions to some specific problems, and there is no sound scientific reason to call for a general ban on these technologies. Each country should have the right to decide whether or not to use these technologies, and to define possible limits of utilisation.

Protection of Intellectual Property

Most countries have decided to privatise their seed industry, including plant breeding activities. As a consequence, this requires incentives to research through protection of intellectual property rights. The plant breeder community is in favour of the strong intellectual property protection necessary to ensure an acceptable return on research investment and to encourage further research efforts. This protection can be provided either through legal or technical tools.

Legal Protection

The seed industry considers that genes sequences associated with traits, gene constructs and biotechnological processes should be patentable. In contrast, DNA sequences not associated with an expressed characteristic, such as Expressed Sequence Tags (ESTs), should not be patentable. Similarly, if protection for a patented process extents to products resulting from that process, the scope of the patent should not be too broad. It is also unanimously agreed in the plant breeder community that alternative approaches to achieve the same product or characteristic should not be considered as infringing the pre-existing patent and should, therefore, be also patentable.

Protection of biotechnological inventions

The seed industry considers that the most appropriate protection of biotechnological inventions is through utility patents, provided, of course, that criteria for patentability are fulfilled. These criteria are (i) novelty, (ii) inventive step (or non-obviousness, depending on the country) and (iii) utility.

Patent laws provide strong protection to the patent holder. Research exemption and farmer's privilege are limited in general. The research exemption is generally limited to an experimental use exemption, allowing checking whether the patented invention performs as expected. In the European Union, it is not clear whether the Directive provides for a true research exemption. In absence of jurisprudence, its interpretation remains unclear. Farmer's privilege exists in some patent laws (e.g. in the EU Directive), but not in others.

Protection of plant breeder's rights and utility patents

In most cases, GMOs are protected by a combination of Plant Breeder's Rights (PBRs) and utility patents: PBRs protecting the genetic background in which the gene construct is introduced, and the utility patent(s) protecting the biotechnological invention(s). This system allows appropriate protection of the legitimate interests of both plant breeders and biotechnologists. Nevertheless, this may raise some potential problems, since two forms of protection with different rights and exemptions then coexist in a single product.

This is particularly true regarding the breeder's exception and the research exemption: in the Convention of the International Union for the Protection of New Varieties of Plants (UPOV), there is a provision allowing free use of protected commercialised varieties for further breeding, the so-called breeder's exception. In utility patent laws, there is a research exemption, frequently limited to an experimental use exemption. A licence from the patent holder is necessary to use a patented invention in a breeding programme. The development of GMOs containing patented inventions may lead, therefore, to a reduction of breeder's freedom to operate. This is the reason why the International Association of Plant Breeders (ASSINSEL) adopted a statement in 1999 on development of new plant varieties and protection of intellectual property. As far as GMOs are concerned, this statement says that:

"When a commercially available plant variety protected by PVP [plant variety protection] contains patented traits, it should remain freely available for further breeding, according to the breeder's exception provided for in the UPOV or UPOV-like systems. ASSINSEL considers that, if a new plant variety, not an essentially derived variety, resulting from that further breeding is free of the pre-existing patented traits, it should be exploited freely. On the contrary, if the new developed variety is an essentially derived variety or if it still contains patented traits and/or patented genetic causative agents, a license from the owner of the initial variety or of the patented traits must be obtained."

The development of GMOs may also have an impact on the so-called farmer's privilege, allowing farmers to conserve and use on their own holding seed saved from their own harvest. The mandatory exception to the PBRs applying in case of non-commercial uses (e.g. the case of subsistence farmers), as provided for in the UPOV Convention, does not exist, or needs to be confirmed by jurisprudence, in patent laws of some countries. Therefore, in these countries, in the case of transgenic varieties, the presence of patented inventions in a variety may affect the "farmer's privilege" provision of the UPOV Convention.

Since it is more and more difficult for breeders to continue their activities without access to biotechnological inventions, and for biotechnologists to exploit their inventions without access to improved genetic backgrounds, there is a trend to develop voluntary cross-licensing agreements between holders of PBRs and patents. This is for the reciprocal benefit of both parties.

Other forms of protection may be used to protect GMOs or biotechnology inventions. One option is to maintain it as a trade secret, whereby the inventor does not publicly disclose any information about the invention; however, the invention is liable to discovery and being copied. Another option is offered by contractual agreements. Contrary to patents, protection through these two systems can last forever, and if the invention is never disclosed, the public will never have access to it.

Technical Protection

Where effective legal intellectual property protection systems do not exist, breeders use, when technically feasible and economically viable, hybrid varieties. This system avoids reproduction of the variety as such through use of farm-saved seed. Hybridisation is an effective protection system, available in several crops, and under development in several others.

GURTs are another alternative to stimulate plant breeding activities, in particular as regards:

- — self-pollinated crops in countries where breeders do not or can hardly receive fair compensation on the use of farm-saved seed by commercial farmers; and
- — cash crops (e.g. cotton, tobacco) and some food crops in countries where plant varieties or biotechnological inventions are not protected.

Depending on whether the plant breeder might be interested in protecting the whole variety genome or a specific trait, he might use either V-GURTs or T-GURTs:

— V-GURTs are systems that would prevent production of farm-saved seeds. As a consequence, the whole protected genome could not be used by farmers for production and use of farm-saved seed.

— In contrast, T-GURTs would only prevent, without paying royalties, re-use of a specific high-added-value gene, such as an insect resistance gene or an output trait. In the meantime, the rest of the genome might be used by farmers for the production and use of farm-saved seed. Only the high-added-value gene protected by T-GURT would not express in the farm-saved seed.

Moreover, it is doubtful that transfer of apomixis (a natural phenomenon occurring in some plant species, whereby plants can produce seeds without fertilisation) to crop species would be fully exploited, if not associated with GURTs. Apomixis would facilitate and speed-up the plant breeding and seed production process. It would also make it easy to produce farm-saved seed and could, therefore, constitute a disincentive to breeders and researchers, if not associated with a technical protection system. GURTs are the perfect complement to allow full use of apomixis in crops. The most acceptable solution would be probably to combine apomixis with T-GURTs that would switch apomixis on for the breeding and seed production processes, and would switch apomixis off after commercialisation.

Today, GURTs are still at the experimental stage. These technologies have not yet been tested in the field. Therefore they are still very far from any commercialisation. It is very difficult to predict what will be the development of GURTs in the medium term. Their development will be influenced by a highly political and emotional debate, while, as for any other new technology, GURTs should be assessed further through a sound, science-based risk-benefit analysis. Then, it is up to each sovereign state to decide whether to use or not use these technologies, and for what purpose.

Application of Biotechnology

Economies in Transition

Public and private sector entities in several developing countries and countries with economies in transition have invested in biotechnology. The vitality of this sector is shown by the establishment of biotech industry associations in countries such as India, Argentina and Brazil, or at continental level, such AfricaBio for sub-Saharan Africa. China is also very actively embarked in plant biotechnology.

Technology transfer

Several companies and public institutions in Western countries are involved in technology transfer to developing countries and countries with economies in transition. They provide licences on most favourable terms, sometimes free. The most relevant organisation is International Service for the Acquisition of Agri-Biotech Applications (ISAAA), an organisation of seed companies and prestigious foundations, that aims to transfer agri-biotech inventions to developing countries. One can mention, among other, an ISAAA project in Kenya to transfer virus resistance to sweet potato varieties. "Golden Rice" (rice with higher vitamin A and iron content to remedy deficiencies in developing countries where rice is the basis food crop) and the International

Rice Genome Sequencing Project are other relevant examples of collaborative efforts to facilitate access of developing countries to modern biotechnology.

Field trials

Field trials of transgenic crops are being carried out on all continents according to BINAS (a UNIDO database) and Bio-Track (an OECD database). In addition to industrialised countries, at least Argentina, Bolivia, Brazil, Bulgaria, the Czech Republic, Egypt, India, Kenya, Mexico, the Philippines, the Russian Federation, South Africa, Thailand and Ukraine are testing GM crops in the field.

As far as Central and Eastern Europe are concerned, the following transgenic crops are being tested:

— *Bulgaria:* alfalfa and tobacco.

— *Czech Republic:* maize, rapeseed and sugar beet.

— *Russian Federation:* maize, potato, soybean and sugar beet.

— *Ukraine:* maize, potato, rapeseed and sugar beet.

Commercial use

According to ISAAA, Argentina is, at the moment, the second-largest grower of transgenic crops. With 6.7 million ha of transgenic crops, Argentina represented approximately 17% of the global transgenic crop area in 1999. China ranked in fourth position with, officially, 0.3 million ha (this figure is an underestimate as it does not include the large acreage of transgenic tobacco cultivated in China). South Africa and Mexico have also initiated commercial use of transgenic crops. Brazil is in a special situation: while transgenic crops are not yet allowed for commercial use, transgenic soybean is grown on about 1 million ha in southern States. Brazilian farmers have gained access to this technology through informal exchange of farm-saved seed with farmers in northern Argentina.

As far as Central and Eastern Europe is concerned, the first transgenic varieties were released in 1999. Herbicide-tolerant maize was grown in Romania on 14 250 ha, and in Bulgaria reportedly on 12 000 ha. *Bt* potato was grown in Ukraine and Romania on limited areas (less than 1 000 ha).

Future Prospects

Developing countries and countries with economies in transition will greatly benefit from biotechnology. This is particularly true for crops that are of limited interest to the seed industry, such as roots and tubers, plantain or pulses, which are basic food crops in the least advanced countries. Tissue culture is already used in some of these countries to speed up multiplication of high quality planting material. Detection and eradication of diseases is also an important challenge for vegetatively propagated crops and pulses, which carry viruses with the propagating material. Use of marker-assisted breeding and transgenesis will also facilitate introduction of important genes that could provide resistance to pests, tolerance to salinity, improved nutritional patterns or even vaccines.

In all other countries, be they developing countries (e.g. Argentina, Brazil, South Africa, China, India) or countries with economies in transition (Eastern and Central Europe), biotechnology is already widely used in plant breeding and planting material propagation, either experimentally or commercially. The role will be rapidly increasing, in particular in countries that produce food for domestic consumption and that will, therefore, be little affected by consumer's acceptance of GMOs in industrialised importing countries.

Role of FAO's

FAO adopted in 2000 a Statement on Biotechnology. This statement recognises that biotechnology provides powerful tools for sustainable development, and can be of significant assistance to increase agricultural production and productivity to meet the needs of an expanding population. It also acknowledges that biotechnology could lead to higher yields on marginal lands, and could improve food quality and the health of many low-income communities. In parallel, FAO recognises the potential risks to human health and the environment. In that respect, the statement emphasizes that caution must be exercised in order to reduce risks. It also re-affirms support to science-based analysis, to objectively determine, on a case-by-case basis, the benefits and risks of each individual GMO.

FAO emphasizes that efforts should be made to ensure that developing countries in general, and resource-poor farmers in particular, benefit more from biotechnology. It is proposed that this be addressed through increased public funding and dialogue between the public and private sectors. The statement also recommends that FAO promote development of networks on plant biotechnology, provide technical information and assistance, as well as analyses and, whenever the need arises, act as an "honest broker" by providing a forum for discussion.

This FAO statement defines clearly the role of the Organisation in assisting its member countries to better benefit from biotechnology. FAO constitutes an appropriate forum for negotiating international regulatory instruments on plant biotechnology. Thus, Codex Alimentarius (a joint FAO/WHO forum) has established a Task Force to develop guiding principles on safety analysis of foods obtained through modern biotechnology; the Commission on Genetic Resources for Food and Agriculture (CGRFA) is discussing a possible Code of Conduct on Biotechnology; and the Interim Commission on Phytosanitary Measures (ICPM) is investigating what should be the role of plant protection organisations as far as the implementation of the Biosafety Protocol is concerned. Additional forums may be established, as needed, to tackle new topics that are not covered by existing forums (such as CBD or OECD).

Development of Cooperation Networks

Development of networks is the most effective way to promote collaboration among countries and stakeholders (governmental organisations and industry) to achieve common objectives. This is of particular interest for biotechnology. Existing crop-related or regional networks could contribute to develop cooperation on R&D in biotechnology. This is probably wiser than establishing new networks dealing specifically with biotechnology. Indeed, we have to keep in mind that biotechnology is a tool and therefore networking activities on biotechnology should be considered as a component of seed or plant breeding networks.

Technical Assistance

Developing countries and countries with economies in transition in general, and the less advanced countries in particular, need technical assistance and advice to develop their national programmes on biotechnology. FAO has gained extensive experience in providing technical assistance to such countries through national or regional field projects and training programmes. These activities should include a component on plant biotechnology, where relevant. FAO plays also a key role in information dissemination. This is also a very relevant role vis-à-vis biotechnology, since technical progress goes very fast, and it is difficult for researchers in developing countries and countries with economies in transition to keep up-to-date on recent scientific developments. Analyses of the regulatory, scientific, technical, trade and political context surrounding development of biotechnology is also very valuable for decision-makers in FAO member countries. FAO could provide, on a regular basis, neutral and objective analyses on biotechnology to help decision-makers take appropriate decisions.

Technology transfer

FAO could act as an adviser or a facilitator between parties involved in technology transfer, in order to ensure that technology reaches the less advanced countries under preferential conditions, taking into account possible legal and trade implications. FAO should ensure that subsistence farmers benefit as well from biotechnology.

Suitable Policy advice

FAO could play a strategic role in advising member countries on suitable policy aimed at promoting appropriate use of biotechnology. This implies development of relevant seed policy. FAO regional networks and consultative groups on seeds could play an important role in that respect, in particular in the perspective of regional collaborative efforts.

9

Sustainable Seed Development Technology

The term seed security suggests a well-defined concept, in practice seed security means very different things for different farmers. For simplicity reasons, reference will be made to the three main classes of agriculture: low (external) input agriculture, the intermediate situation of 'medium input agriculture', and agriculture based on high use of external inputs. It must be realised that this is a great oversimplification, but it is useful here to avoid making the discussion too complex.

In low-input systems, seed security may be achieved when a farmer is capable of producing enough crop so that he/she can reserve some of it as seed for the next planting season. It may be sufficient for him or her to benefit from the gradual genetic change the crops makes in a situation of natural selection, of farmer selection, and by introducing other material once and a while from neighbours in the area. The ideal picture is a steady-state farming system.

Mostly however, the farmer will not produce enough. Not enough to feed his family, for example due to drought, exhausted soils, diseases, etc., or simply not enough because he wants to sell more on the market for cash. Cash which is needed for medical care, education and other services or goods. This means he will be looking for increased yields by introducing 'improved' varieties on his farm. If these are not readily available, this already leads to insecurities on his or her part.

The situation gets worse when he has sold or eaten all of his crop and nothing is left for seed. Then the situation gets precarious and can be considered highly insecure. In large parts of the world, especially in the drier zones of Africa, Latin America, and also Asia, farmers are in exactly this situation: even in the absence of social unrest or natural disasters, farmers have a continuous problem with getting enough seed of the desired kind in time.

In medium-input agricultural systems, where farmers buy agrochemicals, fertilisers and seeds more or less frequently, a steady state could be defined as the situation in which the farmer is able to produce enough seed for himself, to exchange seed with his neighbours, and to buy seed of formal system varieties, whenever he wants. He will normally produce enough crop to sell a sizeable portion of it on the market. In most areas of the developing world, we will see that farmers are frantically looking for new improved varieties, but that the formal system fails to produce adequate numbers of them and certainly not the quantities needed. This means that also the intermediate type of market-oriented farmer is constantly in need of better seed, and that only with great difficulty can he maintain yield levels stable. Although this type of farming has more resilience, here also the situation worsens quickly when social unrest or natural disasters occur.

In high external input agricultural systems, farmers usually sell all of their produce and buy new seed every year or every second or third year. This they can afford because of the regular supply of quality seed by the formal system: by government agencies or private companies. The farmer is relying on their efforts to produce high quality seed, and does not bother to produce it himself because they have specialised equipment, stores, chemicals, packaging material and can do a much better job. Although this system in itself is not seed-insecure, we must realise that the farmer is very dependent on these external sources, and that he cannot easily switch back to seed production of his own.

In this system, there is a constant great need for improvements: higher yielding varieties, more resistance to pests and diseases, tolerance to salinity, but also new drying, storage and packaging technologies, for instance. And here also, the present systems fail to produce these innovations at a satisfactory pace. The farmer is thrown back onto his own resources as soon as turmoil starts, since production and distribution activities of the formal sector will then soon cease. Linked with genetic diversity, an important element in seed security is the access to a diverse set of genes: they help the farmer to grow crops successfully in a variable environment, which may be variable in space as well as in time.

Farmers must reduce risks of crop failure as much as possible. In low and medium-input systems they usually grow a number of varieties of one particular crop, not just the most high-yielding. Different varieties are needed for the different conditions encountered in the field (variation in space), erratic climatic behaviour (variation in time), and for the reduction of the chances for pests and diseases. Moreover, different varieties often have different uses, thus making them de facto different crops.

Seed Security in Peacetime

Participatory Approach of Plant Breeding

One of the better ways to assist farmer innovation seems to be the participatory approach of plant breeding. The combined action of trained breeders and experienced farmers has the potential of breeding well-adapted varieties which will be available without delay, and which will find their way basically through the informal system.

Participatory Seed Technology Development

One very crucial step in crop production is seed production, including storage. Annually, enormous amounts of seed are getting spoilt or lost during post-harvest operations. At times, losses are well over 50%. The seeds get lost during the processes of harvesting, drying and cleaning, while the physiological quality of seeds rapidly declines if no adequate precautions are taken.

Formal systems of seed production have been set up to produce seeds centrally. Due to all kinds of reasons, these systems often fail. The number of varieties is too restricted, the seeds are not available at the right time, nor in the desired quantity or quality, etc. At present, many local initiatives are therefore being taken, mainly by NGOs, to solve this problem. Local small initiatives are being set up, farmers are encouraged to combine efforts, to engage in replicas of the formal system at a local scale. Many NGOs, however, lack the specific knowledge which is needed for seed production, and this often results in disappointing results.

Like with plant breeding, a more participatory approach is needed. It must be taken for a fact that most low-input farmers will continue to produce their own seed for a long time to come, most probably forever. The formal model of seed production is simply not adequate for this category of farmers and must, therefore, entirely be abandoned for this group. Only those initiatives must be taken which have been identified by farmer and researcher/facilitator together. This means that if a farmer prefers to produce seed for himself alone, he must no longer be urged to enter into co-operatives, because it worked so well in Western Europe. There are not so many economies of scale when you are a small farmer and labour is not the limiting factor.

Together with the farmer, the most optimum way of seed selection must be developed, and together the harvesting and storage technologies must be reviewed. It is likely that the researcher can learn from the farmers practices, and with his theoretic background can probably explain to the farmer why these traditional ways make sense. It is likely that these practices can be improved without too much effort: the farmer must be made aware of the critical steps, and their interaction. If for instance, the farmer decides to burrow his seed to keep it away from thieves, he must be warned about the dangers of moulding. It must be explained to him that he should dry his seed more rigorously than when he would store it in a storage barn. When he then puts the seeds into a moisture proof container, such as a metal can with lid or an oil bottle with screw cap, he can bury his seed safely.

New materials may be developed and distributed specifically for small farmer use, like simple multi-use plastic containers for seed storage, or special simple procedures for improved methods of seed germination. Also, local indigenous knowledge learned in one area may be transferred to another area, where farmers will readily adopt it when the technology is properly introduced. In this way, storage of seed using specific natural disinfectants and other methods to protect seeds from mould and insects can be introduced on wider scale.

The intermediate type of farmer must be assisted in a similar way: first identify together what is needed, and not bring what they should use in our opinion. Intermediate technologies provided the manufacture is entirely local, remain a very important aspect of agricultural development for this large group of farmers. For high input farmers, the formal system should be strengthened: donors have massively abandoned this sector, leaving the system in disarray. The knowledge present in this system, however, is not yet lost and it is very high time that this important sector of the countries' economies get new support.

The NARS involved must be taught to work totally and exclusively in a client-oriented manner. They should start working together with farmers in the development of new breeding criteria and farmers should become heavily involved in the decision-making process. Farmers' committees should visit the trials several times per year, they should be involved in the initial selections as well as the later ones, they should have a say in how many varieties should be produced and released, and they should indicate why and how more diversity could be created. This involves a completely different approach, and a very important one.

Donor Inertia and Privatisation

Donors have been frustrated by the little impact their support has had. It was suggested that the inefficient governmental organisations and parastatals should no longer be supported. It was suggested that they should privatise or that the private sector would come in by itself and take over the initiative. Nothing of the kind has happened in the poorer countries of the world. Proponents of the free market have made a grave miscalculation, which now has led to the destruction of the formal system.

In most West and Central African countries, for instance, the formal production and quality control of seeds has stopped. Privatisation of governmental agencies or the emergence of a private sector has not occurred because in a society without money, no money can be earned. This principle is even more true for seeds: they are not scarce in the economical sense, they are not irreplaceable, because farmers have access to alternative sources. This means that even if the farmer has some money, and even if the value-for-money-ratio is favourable, they may still decide to buy food which they do not produce themselves or medicines for which no alternatives exist.

Most farmers grow varieties developed and released through this system, certainly the high and medium-input farmers. Although they seldom buy new seed, the varieties they grow have been acquired via the informal exchange and farmer production system. And all these farmers are waiting for improved varieties to come, but they have to wait for far too long. It is time to stop the negativistic thinking, and to take a fresh new approach: short-term scoring of projects is not in the interest of the recipients, the sustainability of assistance should be stressed.

This means that so long as sustainability has not been achieved, support will have to continue. It is irresponsible behaviour to come into a country, overthrow what is there and to leave when progress seems to halt. Countries have the right and should try to establish a strong local system. Regional initiatives will prove be very valuable, and should be developed further along the lines of client orientation and participation, but donors should be there to support the local systems, not to replace them.

Donors must have the courage to engage into long-term commitments, which means that donors should support twinning between institutes and agencies from North and South. This also means that they should make choices: not every agency can change the world on its own. That may also mean that not just the cheapest solution is the best. These longer-term relations, if successful, go beyond the technical level and open possibilities for many more initiatives. They should have a programme-approach: they should be flexible and have possibilities for new initiatives which develop from the interaction between partners and which can only emerge along the line. In the end, this kind of co-operation will prove much more cost-effective than the hit-and-run projects of today.

Seed Security in Times of Privation

In former aid campaigns, it has become clear that seed aid with seeds of crops or varieties which are not known in the area leads to very disappointing results. The crop may fail in an early stage, but it may also be that the farmer invests a lot of energy in a crop which matures too late, leaving him with no harvest at all. Another danger is that because of the utilisation of exotic varieties, the local stock of varieties will be even further reduced. It is of imperative importance, therefore, to include the right (range of) varieties in aid programmes.

Then the question arises of where to obtain this material. Since most grains can be used as seeds, one of the best options is to buy locally produced grain and distribute it as seed. Apart from the negative effect on the food reserves and their price, it is much preferred over other alternatives. In order to avoid failure during sowing, it is important that this material is checked for its germination capabilities. Under the circumstances, germination percentages of as low as 50-60 % may still be acceptable, as long as the material is of the desired variety. It is much more important to provide the correct varieties than the optimum germination capacity: the crop should not only be able to germinate, but also be able to grow and flower successfully and produce a mature crop which can be harvested.

This requires some seed expertise usually unavailable to relief agencies. In order to be able to buy the correct grain for seed purposes, one must be absolutely sure about the correctness of the varieties. Grain merchants will otherwise be tempted to mix the seed with less valuable ones, or sell totally unacceptable qualities and varieties in order to fetch the high prices which are being paid for local material in emergency situations.

If grain from other areas is brought in for food purposes and not for seed, it must be realised that in the absence of seeds in the area, this grain will be used as seed. Such cases must be avoided, since the farmer may have unfounded expectations, and all his work will be in vain. In such cases, it is better to send food aid exclusively in the form of flour, so that the farmer will keep on searching for alternative seeds. Of which he may find a few indeed: it has been shown that even in the most desperate circumstances, some original local seed stocks could still be found. The farmer may obtain perhaps a small quantity, as social favour or in return of labour. For him it is better to sow less, than to have a complete crop failure from an exotic variety.

One of the major problems is that seed can not be produced at a higher rate that its natural multiplication rate. This means that assuming we can grow the variety in another area, that in order to produce a ton of seed, approximately 20 kg of rice, 10 kg of maize, or 7 kg of sorghum are needed. These figures are based on average seed yields, but yields will be much less if the seed is produced close to the disaster area. This means that it would be necessary to create and maintain a stock of at least a few hundred kg of each variety.

This will need a coherent effort of FAO, NARS and CG centres, in order to maintain sufficient duplicates and kilograms around the world. The seed stocks should preferably be kept at sub-zero temperatures, because multiplication may not be possible in the areas where the collaborators have their facilities. If no large quantities can be kept and much smaller quantities will be possible, this will mean that sizeable portions of seed will become available only after two seasons.

An emergency production system involving complementary areas on the Northern and Southern hemisphere with similar climates (if existing) would in part solve this problem. These areas should be tested for their suitability to produce the various varieties in stock. For quick and effective initiatives, it is also imperative to know exactly what seeds will be needed. At present, there is no comprehensive inventory of the various varieties used in all candidate countries. A first attempt must be made to make this inventory.

The quickest way would be to invite agencies which are active in the area to produce these data. A central database could be developed to store these data, to update them regularly and to release them in a form usable for relief agencies. Apart from the location of the various seed samples, this database should also contain the addresses of all known sources of seeds in the area, including seed companies, community seed banks, prominent farmers who are known for their seed, and further data on the main agricultural requirements of the varieties, their main uses, etc. Because of ownership issues (farmer's rights), some reluctance to collaborate may be encountered in peacetime.

Functions of the Different Actors

Role of Various Players

In order to work client-oriented and with the involvement of these clients, it follows that a different approach is needed by the various actors involved in seed development. The list is by no means exhaustive, and if a feature appears it does not mean that this is not yet done anywhere.

Relief agencies

— develop knowledge on varieties for the various environments;

— develop skills to assess seed quality and to evaluate grain to be used for seed; and

— develop a system to produce large quantities of seed in a short period of time.

Role of CG Centres

— establish more regional offices, to acquire more insight in local needs;

— generate larger amounts of genetic material: put provision with useful genes before formal system requirements;

- — assist NARS in their new tasks by training in new methodologies;
- — develop GIS-information on local use of varieties; and
- — organise the regional collection and characterisation of local varieties together with NARS.

Role of governments

- — stimulate local research and research collaboration;
- — invest in the seed sector;
- — review the milestones for development; and
- — develop new and less stringent seed regulations.

Role of NARS

- — end the practice of keeping genetic diversity beyond the farmer's reach;
- — abandon formal thinking and develop participatory skills;
- — concentrate on assistance for small farmers;
- — improve seed production capabilities for high-input agriculture;
- — collect field data about varieties used and store them in an accessible way; and
- — assist in the collection of seed for possible future relief activities.

Role of Private Sector

- — develop varieties using combinations of promising lines with local varieties; and
- — collaborate with NARS.

Role of NGOs

— ask assistance from NARS and other seed experts with all seed activities: especially at the planning stage;

— develop both community and individual farmers initiatives; and

— involve NARS in your activities: expose them to participatory methodologies.

Role UN-agencies, FAO

— take care of global and regional issues; and

— support the CG-NARS system.

Role of Multilateral and Bilateral Donors

— invest in the seed sector;

— make in long-term commitments using a programme approach; and

— promote North-South relationships: encourage collaboration.

Seed Supply

Seed supply in developing countries relies on two distinct sources:

— the seed industry, be it public or private, and

— on-farm seed, which is the main source, particularly in the most remote areas.

Seed Industry

It is difficult to collect data on the commercial seed market in developing countries. However, it is possible to say that the commercial market represents less than 10% of the seed consumption, in particular for the staple food crops. This is due to several factors, but the most important one is certainly the weakness of the seed distribution networks. Due to the lack of involvement of the private sector, the number of seed suppliers is not sufficient. In some countries, seed production and distribution is a public sector monopoly.

On-farm Seed Production

On-farm seed production is the main source of seed in developing countries, in particular for small-scale farmers (SSF) in the most remote areas. For some food crops such as sorghum, millet, cowpea, cassava and sweet potato, on-farm seed or planting material represents more than 90% of seeds used by the farmers.

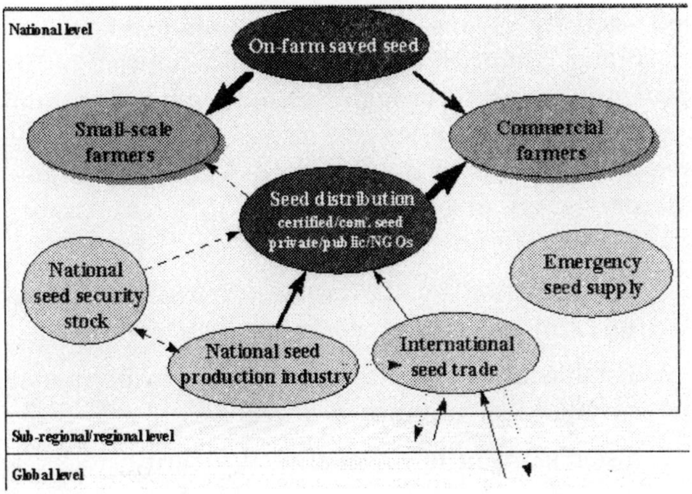

Figure 1. Seed Supply in Developing Countires Normal Year

It must be noted that this percentage is an average, since figures vary a lot according to climatic conditions: after droughts, farmers have no or little on-farm seed available and, therefore, require access to commercial seed or seed provided through relief operations.

On-farm seed must be considered as a major component of seed supply and seed security in developing countries, but it is not a stable one. Consequently, it is difficult and very risky to develop a sustainable seed security strategy based only on such a component.

The two following diagrams show (i) the evolution of seed sources depending on the years (normal year vs. year with calamity disrupting national production) and (ii) that the 'informal seed sector' is a weak component to ensure sustainable seed security:

Integration of Security Strategies

Seed security is one of the most important issues in developing countries in particular in countries or areas frequently subject to droughts or other natural or human disasters. Many countries and international Organisations have already tried to develop seed security strategies at different levels of integration. The different strategies developed up to now are:

1. Conservation by the farmers themselves of small on-farm saved seed stocks for their own needs;
2. Establishment of small seed banks to be managed at village or community level;
3. Establishment of national seed security stocks; and
4. Development of sub-regional or regional seed security networks.

Options 1 and 2 function well in the case of a normal year, but usually fail in the case of disaster. Option 3 aimed at establishing national seed security stocks is a basic strategy in drought-prone countries, but stocks should not exceed 20-25% of the annual seed requirements due to high costs to conserve such stocks. Option 4 aimed at developing seed security networks in order to facilitate co-operation among countries in providing farmers with good quality seeds of suitable improved varieties, is certainly the best solution. Indeed, it seems obvious that achieving seed security at sub-regional or regional level is more efficient and cost-effective than at national or local level.

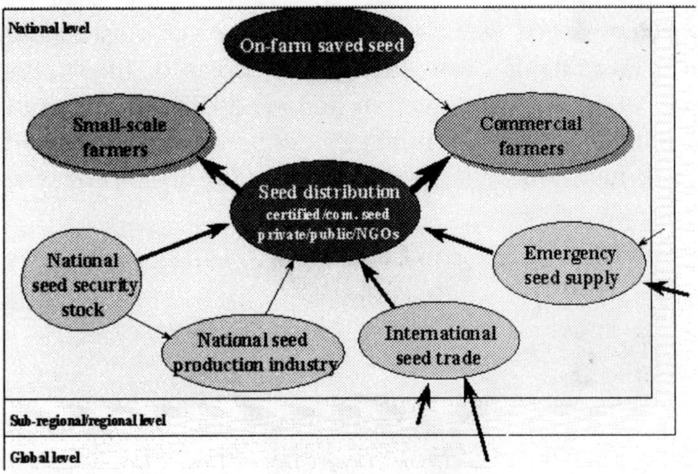

Figure 2. Seed Supply in Developing Countires Year with Calamity

An indirect objective of seed security schemes is also to limit seed donations as carried out currently by many bilateral or multilateral relief programs. Indeed, it has to be noted that in the case of emergency supply, farmers are frequently provided with seeds of poor quality (as regards germination or health) or of varieties which are not adapted to their area. In addition, such emergency supply can have adverse effects on the long-term development (i) through introduction of new pests or diseases or of new genotypes which will make crosses with local varieties or landraces and (ii) in disturbing or destroying existing commercial seed distribution networks. A good alternative would be to modify seed relief programs in order to promote sustainable private seed sector by distributing seed vouchers instead of seeds.

Sharing Responsibilities

In the future, the role of the seed industry and seed trade to achieve seed security at regional or sub-regional level on a sustainable basis should be increased. International seed trade, seed production and seed distribution should be mainly the responsibility of the private sector (whether small, medium or large-size companies or co-operatives).

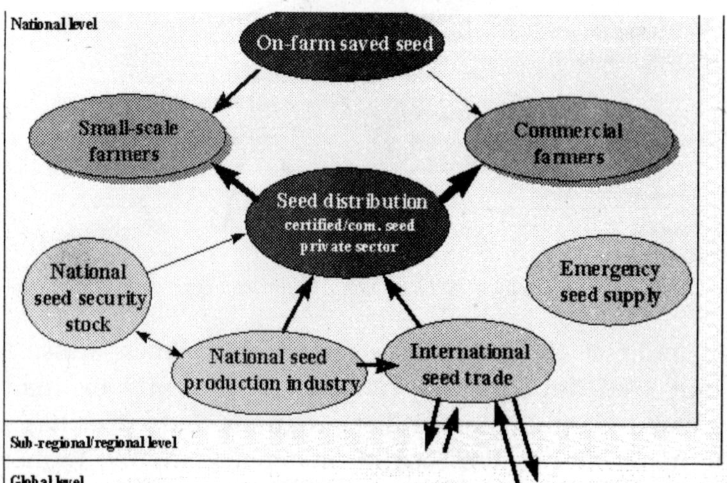

Figure 3. Seed Supply in Developing Countires Expected Situation

Activities of seed development projects, extension services, NGOs, etc. should focus on the development of seed distribution networks based on private retailers and co-operatives, in particular in the most difficult and remote areas. Seed security strategies should be designed at sub-regional, regional or global level, thus involving the seed industry of neighbouring countries. Pursuant to that objective, barriers to international seed trade should be removed. The expected situation is summarised in Figure 3.

In such a perspective, the role of national authorities should be to facilitate international seed trade and harmonise national legislation among developing countries of the same region with the assistance of appropriate international organisations.

Barriers to Trade Seed and Achieve Seed Security

Within the different regions or sub-regions, several barriers to seed trade have to be removed, in order to facilitate exchange of commercial seeds among and within the countries and in order to implement a strategy aimed at achieving seed security at sub-regional, regional or global level. The main obstacles are currently:

— *The weakness of the seed distribution networks:* it is certainly one of the weakest component of the seed trade in developing countries if we want that quality seed of improved varieties reach SSF. Political strategies should have as an objective to strengthen seed distribution networks. Where possible and appropriate, projects, NGOs and extension services should teach SSF how to enter the formal seed market and realise substantial benefits in establishing small- or medium-size private companies or co-operatives to be involved in seed distribution. It would surely ensure sustainability when projects end.

— *Variety release procedures:* most developing countries have a list of authorised varieties, with very long, and sometimes subjective, release and registration procedures. This list is frequently limited to a very few varieties developed mainly by the national agricultural research institute or some international agricultural research centres. Many countries do not allow sale of seeds of non-listed varieties, even if better than registered ones. Simplified registration procedures or voluntary registration would facilitate the flow of improved varieties to farmers.

— *Seed certification procedures:* these procedures limit exchange of seeds among countries having different procedures and standards. Two options would facilitate seed trade at sub-regional level: (i) to render seed certification voluntary or (ii) to harmonise procedures, rules and standards in becoming member of international organisations in charge of harmonisation or (iii) a combination of both.

— *Lack of international seed trade rules:* only few national seed trade associations and seed companies of developing countries are member of FIS and apply FIS rules and usage for seed trade, which are the basis for seed trade at international level.

— *Lack of protection of intellectual property:* many developing countries signed the GATT (General Agreement on Tariffs and Trade) and therefore the TRIPs (Trade-Related Aspects on Intellectual Property Rights) Agreement. According to the TRIPs Agreement, all countries have to develop by 2005 a national legislation regarding protection of new varieties of plants either by patent or by a sui generis system such as the UPOV (International

Union for the Protection of New Varieties of Plants) Convention. Nevertheless, up to now, only a very few developing countries are member of UPOV. This is a main obstacle to have access to improved varieties bred abroad.

— *Unreasonable phytosanitary requirements:* companies wish to be allowed to import and export healthy seeds. They recognise the need for clear and scientifically based phytosanitary regulations. Nevertheless, it must be noted that phytosanitary regulations are currently one of the main obstacles to international seed trade. Therefore, there is a need to harmonise regulations among countries, in particular at regional level. The IPPC (International Plant Protection Convention) provides the frame for such an harmonisation in order to avoid unreasonable and unfair trade barriers.

Sustainable Strategies for Promoting Seed Security in Disasters

The problems caused by major disasters are of considerable weight and complexity. Schell noted that a disaster arises when an entire population, or only a part thereof, is exposed to an occurrence which directly threatens human life or causes such a degree of damage to the economic and social structures that survival becomes impossible.

The community will strive, using every means, to overcome such emergency situations. However, Webb indicated that the number of individuals affected by such events is growing at a rate of 16% every year. The figures have, therefore, increased from 44 million in 1980 to an excess of 175 million in 1993. There are natural disasters caused by drought, floods, fires, earthquakes, volcanic eruptions, etc., which occur and will continue to occur in the future with a certain degree of regularity. In addition, there are disasters caused by man which result in the destruction and elimination of the entire political and social structure and which lead to the migration of great masses of population.

The number of refugees and those who are forced to leave their homes due to conflict is growing at a frightening rate. While the figures for such displaced people fluctuated annually in 1960-75 between 2.5 and 3 million, this figure reached 38.8 million in 1990-95 with over a half of those involved being from the Sub-Saharan African Region. Disasters are further overburdened when the effects of drought are compounded by the effects of conflict (Table 1). This particular phenomenon arose between 1979 and 1995 in the African countries of Burundi, Eritrea, Ethiopia, Djibouti, Rwanda, Somalia, Sudan, Tanzania and Uganda.

The majority of the people affected by these events are living today in conditions of chronic starvation and food shortage. As a result of the spread of conflict and increased occurrence of disasters, there has also been a marked increase in the burden of providing assistance and each year the sums spent on humanitarian aid in such disaster areas are multiplying.

In general, three specific types of humanitarian intervention may be identified: food aid programmes, food aid projects and emergency food aid. It is, however, necessary today, independently of the method of intervention, to ensure that the availability of food is not only dependent on food aid. There is, in fact, an urgent need to devise other technical, political and economical strategies to encourage, in the countries affected by different types of disasters, the resumption of agricultural activity.

The provision of aid to farmers is provided for in the 20 most important actions listed in the Global Plan of Action for the Conservation and Sustainable Utilisation of Plant Genetic Resources for Food and Agriculture. The aim is the restoration of the suspended agricultural system by providing the farmers with seeds of local varieties which are well suited to the environment. Indeed, the availability of food is closely connected with the availability of seeds. When the seeds are chosen in an appropriate way and distributed with care, they are an important tool in the strategy to ensure the protection and resumption of agricultural activities following every serious emergency.

In each case, it is important to take into account the nature of the disaster involved as this conditions the type of intervention required. With regard to cases of armed conflict, natural disasters and new settlements, the Global Plan of Action provides a brief overview of the interventions and initiatives applied following a number of disasters which occurred in the last ten years. These interventions and initiatives include: the Seeds of Hope Project; the Second Seeds of Hope Project; the Programme on the Re-introduction of Germplasm for Diversification of the Genetic Base of Somali Agriculture Production, CINS-IPGRI; the SADC-ICRISAT Emergency Response Project, South Africa; the West Asia and North Africa (WANA) Network, and the IRRI Cambodia Programme.

Other recent enterprises testify to the increased awareness of the importance of the availability of seeds. These initiatives include a series of seminars and studies and a number of meetings and conferences. In the light of the above, the role of the institutions involved in ensuring the availability of seeds may not be ignored. We should also remember the large international banks for the preservation of germplasm together with the local deposits for germplasm which are promoted in rural areas by NGOs and farmer organisations.

Besides, it should not be forgotten that farmers, particularly those from risk areas, rely on a constant supply of seed rather than on the quality of such seed. As a result they sow the widest range of local species and land races. In this way farmers ensure their own survival and they also contribute to the increase in the genetic pool of the land races and to the maintenance of a high degree of biodiversity. It is therefore opportune that farmers are called upon to play an active role in every aspect: from germplasm conservation to breeding and seed production.

Among the interventions necessary to ensure the socio-economic recovery of the area affected by a disaster, the re-establishment of food independence is of primary importance. This objective may only be achieved if the population involved have adequate agricultural resources. The seed security is the agricultural resource which requires most care due to its complexity and its effects on the social ecology.

Issues of Seed Security

An inward look into the seed supply arrangement and factors that often disrupt these supply systems will be necessary in order to gain an insight into what issues are involved.

Seed Supply Systems

The recognised seed delivery systems fall into two categories - the formal and informal.

Formal seed supply

This supply system generally consists of a research

institution (mainly public), private seed production and marketing agencies and seed quality control organisations. In spite of reasonable successful seed programmes in many developing countries, the formal seed industry arrangements is reported to cater for less than 10% of the seed needs of the farmers. The most important reason given for this is the fact that seed programmes have concentrated on major food and cash crops which are considered national priorities in terms of foreign exchange earning and selected staple food requirements.

Informal seed supply

The informal seed supply system has its roots in the age-old tradition used by farmers to ensure the supply of more than 90% of the planting materials of staple food crops required to meet food security. This comprises mainly of on-farm production, selection and saving of grain, tuber or stalk from harvested crop as seed for the next cropping season. It also includes on-farm seed production surpluses, which are traded in the appropriate local markets in various forms (cash, barter or kind).

The role of this system in germplasm conservation has been well documented. Among the many qualities which made on-farm seed production a credible option for seed security of staple food crops are:

- — Broader national coverage;
- — Maintenance and sustenance of many crops, which are not patronised by the formal seed supply system but which are vital to the survival of the majority of the people, particularly the under privileged;
- — Reasonable and affordable seed costs to other farmers and availability of alternate arrangements for seed payment;

— Inclusion of farmer's varietal preference in the production programme;
— Closeness of seed supply sources to cultivators; and
— Stability of crop yield through the use of genetically broad-based varieties or ecotypes.

Factors affecting Seed Supply

Both the formal and informal seed supply systems are often hampered or disrupted by a number of factors. These include both the abiotic and biotic factors.

Abiotic factors

These factors often determine the extent of success or failure of a seed supply system. These include many acts of nature such as drought, flood, freeze, volcano, etc.

Biotic factors

A number of basic factors are known to disrupt seed supply form time to time. These include pest and diseases as well as activities of man.

Generally the above mentioned factors affect the seed supply of the informal seed sector more than its formal counterpart where there is always a steady infusion of funds and technology to contain on-toward effects. Risk factors identified with on-farm seed production include drought, infrastructure, and hygiene of planting materials, quality of seed, pricing structure and lack of credit support systems.

As reviewed by FAO, a number of risks are associated with on-farm seed production. These include:

— Drought;
— Limited operational capacity of NGOs; and
— Price disincentives.

10

Structural Variations

Diversity is great in the internal and external structure of seeds. This diversity is related primarily to the wide range of methods of dispersal and germination. The morphological variations include differences in size, form, texture, color, and presence or absence of special structures such as arils of exostomic (elaiosome, caruncle), raphal (strophiole), or funicular origin. Anatomical variations include the presence, absence, or position of the storage tissues; the size, form, and position of the embryo; the form and size of the chalazal zone, and the structure of the seedcoat.

The size and position of the embryo vary among seeds. The area occupied by the embryo is inverse to that used by the endosperm, perisperm, or both. A larger embryo has a smaller endosperm or perisperm, and vice versa. The classification system is based on the size, position, and form of the embryo. It represents a modification of the classification system proposed by Martin. Although some types overlap, the taxonomic usefulness of the classification is maintained. Types must be identified immediately after fruit dehiscence, especially in seeds with no maturation drying where embryo development is continuous. For example, in

Virola koschnyi seeds, the foliaceous, haustorial, and divergent cotyledons begin to develop inside the seed after seed dispersal, and because space is limited the cotyledonar blade folds. In *Compsoneura sprucei* the cotyledonar development is faster and the foldings are more conspicuous.

Classification System of Seed Embryos

The following classification system divides seed embryos into three primary types: basal, peripheral, and axile. Basal is further divided into four subtypes and axile into seven.

Basal Embryos

Embryos are usually small, nonperipheral, and restricted to the inferior half of the seed, except in the lateral type. The seeds are medium to large, with abundant endosperm, starchy or oily. The rudimentary and broad subtypes are found in the monocotyledons and dicotyledons; the capitate and lateral subtypes are typical of the monocotyledons.

(1) *Rudimentary:* The embryo is small and globular to ovaloblong. Cotyledons are rudimentary or obscure, sometimes evident, simulating miniatures of the linear or spatulate types; e.g., *Ilex skutchii* (Aquifoliaceae), *Calatola costaricensis* (Icacinaceae), Magnolia poasana (Magnoliaceae), Compsoneura sprucei, Myristica fragrans, Otoba, Virola (Myristicaceae), Minquartia *guianensis* (Olacaceae).\

2) *Broad:* The embryo is as wide or wider than it is high, and very fleshy. It could be globular or lenticular, peripheral or nearly so. This type is common in dicotyledons and monocotyledons that are quite primitive, aquatic, marshy, or parasitic.

(3) *Capitate:* The embryo expands distally into a head-like form (turbinate); e.g., *Tradescantia* (Commelinaceae), *Dioscorea* (Dioscoreaceae).

(4) *Lateral:* The embryo is basal-lateral or lateral, inclined to expand in the plane of the periphery. The embryo is usually less than one-half of the seed (inferior half) but can be larger. Typical in Poaceae.

Peripheral Embryos

Embryos are usually elongated and large. The embryo occu-pies one-quarter to three-fourths of the seed. It is partially contiguous to the seedcoat and often curved, central or lateral, with cotyledons narrow or expanded. The endosperm or perisperm is starchy. As is typical in dicotyledons, one cotyledon can be smaller or abortive (*anisocotyly*); e.g., *Pereskia* (Cactaceae).

Axile Embryos

Embryos range from small (occupying only part of the seed's lumen) to large (occupying the whole lumen), central (axile), straight, curved, coiled, bent, or folded. The endosperm can be oily or starchy. Found in gymnosperms, dicotyledons, and monocotyledons.

1) *Linear:* The embryo is usually several times longer than it is wide. It can be straight, curved, or coiled. The cotyledons are small, not expanded or not developed. Seeds are not minute. Examples include *Annona* (Annonaceae), *Tovomita, Tovomitopsis,* Garcinia intermedia (Clusiaceae), Elaeocarpus, Sloanea (Elaeocarpaceae), Muntingia calabura (Tiliaceae), Garrya laurifolia (Garryaceae), Bertholletia excelsa, Eschweilera, Lecythis (Lecythidaceae), *Ardisia brenesii* (Myrsinaceae), and *Zizyphus* (Rhamnaceae).

2) *Miniature:* Embryos stocky or minute in small-to-minute seeds. Seedcoats are delicate, often reticulate; the endosperm is not starchy. Two types exist:

a) *Dwarf:* The embryo is variable in relative size, small to total (whole seed lumen). It is commonly oval, elliptic, or oblong, with incipient cotyledon development. The embryo is frequently stocky. Seeds are small at 0.3 to 2.0 mm long; e.g., *Clethra* (Clethraceae), *Buddleja americana* (Loganiaceae), several Ericaceae.

b) *Micro:* The embryo is minute or total (whole seed lumen), globular, and usually undifferentiated. Seeds are small, commonly less than 0.2 mm long; e.g., Orchidaceae.

(3) *Foliate:* The embryo is large, one-fourth to all of the seed lumen, central rather than peripheral. Cotyledons are expanded and the endosperm is not starchy. Seeds are medium to large. Four types exist:

(a) *Spatulate:* The embryo is straight, and cotyledons are variable, from thin to thick and from expanded to broad; e.g., *Aspidosperma, Tabernaemontana* (Apocynaceae), *Bixa orellana* (Bixaceae), *Cornus disciflora* DC. (Cornaceae), *Diospyros* (Ebenaceae), *Hyeronima, Sapium* (Euphorbiaceae), *Casearia, Xylosma* (Flacourtiaceae), Cedrela, Melia azedarach (Meliaceae), Genipa americana, Simira maxonii, Psychotria (Rubiaceae), Zanthoxylum kellermanii P. Wilson (Rutaceae), Manilkara zapota, Chrysophyllum cainito (Sapotaceae), Goethalsia meiantha (Tiliaceae), *Guajacum sanctum* (Zygophyllaceae).

b) *Bent:* The embryo is bent and spatulate and usually has thick cotyledons, planoconvex, bent on the hypocotyl. The latter is partially immersed in the area of cotyledon overlapping. In the most common type, the curvature occurs in the plane of the cotyledons [(o=), accumbent, pleurorhizal]. The curvature also can be against the cotyledonar plane (o I), incumbent, notorhizal; or can be oblique (o //) if it is intermediate between accumbent and incumbent. The fourth subtype is the orthoplocal: the cotyledons are incumbent and conduplicate (o >>) as seen in *Avicennia.* The spirolobal embryos (o I I) have incumbent cotyledons, folded once, while diplecolobal embryos have two or more folds (o I I I). The convolute cotyledons are present in *Terminalia.* Examples include *Acer pseudoplatanus* (accumbent, Aceraceae), *Ceiba pentandra* (bent, Bombacaceae), *Crataeva religiosa* (spirolobal, Capparidaceae), *Caryocar costaricense* (accumbent, Cary-ocaraceae), *Sapindus saponaria* (Sapindaceae), *Ficus, Maclura tinctoria* (bent, Moraceae), and *Myrcianthes fragrans* (accumbent, Myrtaceae).

c) *Folded:* The embryo has thin cotyledons, expanded and folded in several ways. Embryos with conduplicate cotyledons are frequent in Sapindales and Malvales. In Vochysiaceae, the cotyledons are foliaceous, large, and convolute as seen in *Vochysia, Qualea,* and *Calycanthus occidentalis* (Calycanthaceae). Examples of folded embryos include *Bursera simaruba* (L.) Sarg. (Burseraceae), *Couratari, Couroupita* (plicate; Lecythidaceae), *Tilia americana*

(Tiliaceae), *Myrcia splendens* (conduplicate; Myrtaceae), and *Pimenta guatemalensis* (involute; Myrtaceae).

d) *Investing:* The embryo is erect; the cotyledons are thick, overlapping, and encase a small hypocotyl. The endosperm is missing or very reduced. The embryo may be confused with bent or spatulate embryos. It can be associated with cryptocotylar seedlings. Examples include *Alnus acuminata* (spatulate-investing; Betulaceae), Calophyllum brasiliense (Clusiaceae), Prioria *copaifera* (Fabaceae-Caesalpinioideae), *Pentaclethra macroloba* (Fabaceae-Mimosoideae), *Dipteryx oleifera* (Fabaceae-Papilionoideae), *Quercus* (Fagaceae), *Ocotea, Nectandra* (Lauraceae), *Carapa guianensis, Guarea* (Meliaceae), *Nephelium lappaceum, Cupania* (Sapindaceae), and *Quassia amara* (Simaroubaceae). The total or partial fusion of the cotyledons along the adaxial surfaces is common in seeds with cryptocotylar germination such as *Calophyllum, Carapa,* and *Guarea.*

Seed Structure and Storage Behaviour

Some seeds lose water during maturation drying and gradually acquire tolerance to desiccation, while others maintain a high water content, do not experience a reduced cellular metabolism, and are sensitive to desiccation and temperature decreases. The tolerance/ intolerance to desiccation shown by the seeds in their natural environment is also exhibited when they are stored.

Sensitivity to desiccation limits a seed's storage potential, genetic conservation, and use in trade. Roberts defined two types of seeds based on sensitivity to desiccation: orthodox seeds (those that undergo maturation drying) and recalcitrant seeds (those that do not undergo maturation drying).

The ability to tolerate desiccation by orthodox seeds is associated with metabolic changes such as respiration decrease, increase of some carbohydrates or oligosaccharides, and accumulation of dehydrines (LEA proteins). During germination the seeds lose this tolerance, frequently several hours after radicle protrusion. Dehydration at this stage leads to irreversible damage, in which the peroxidation of lipids and free radicals has an important role. However, mature orthodox seeds can be dehydrated without damage to very low levels of moisture (1 to 5 percent) and in a variety of conditions. Bound water (structural) is less easily frozen than free water. Bound water seems to be a crucial component to tolerating desiccation, and in the orthodox seeds all water is bounded. In storage, the longevity of seeds increases with a reduction of the water content in a predictable and quantifiable manner. Bonner and Vozzo subdivide these seeds into (a) true orthodox, which "can be stored for long periods at seed moisture contents of 5 to 10 percent and subfreezing temperatures" and (b) suborthodox, which "can be stored under the same conditions, but for shorter periods due to high lipid content or thin seedcoats."

Recalcitrant seeds are rich in free water and neither tolerate nor survive desiccation. They die when the water potential reaches levels similar to those of permanent withering in many growing tissues (-1,5 to 5,0). The dehydration of intolerant tissues (recalcitrant) provokes membrane deterioration (plasmalema and mitochondria), protein denaturalization, and reduction of both the respiratory rate and the ATP level. The oxidative processes and the free radical seem to be involved in cellular and molecular deterioration. The seeds show a strong resistance to rehydration and the loss of cellular integrity leads to a loss of viability. Recalcitrant seeds are present in at least 70 percent of tropical trees. Their sensitivity to temperature lowering is due to the high water content.

Because some seeds did not fit readily into either orthodox or recalcitrant categories, Ellis and others proposed a third category: the intermediate seeds. These seeds survive desiccation at intermediate moisture levels but not to the degree of orthodox seeds. This last category can be considered arbitrary, and the existence of a recalcitrance gradient throughout the different species has been suggested.

Variations in the sites of water storage and the gradual damage observed in the seed's tissues when it dehydrates, can be illustrated by the seeds of the following species: *Calophyllum brasiliense* (Clusiaceae), *Otoba novogranatensis* (Myristicaceae), *Minquartia guianensis* (Olacaceae), *Caryocar costaricense* (Caryocaraceae), and *Lecythis ampla* (Lecythidaceae) (Table 1).

How does dehydration affect seeds? In *Calophyllum brasiliense* the seedcoat is hard and resists desiccation; water loss is slow. Once the seedcoat dehydrates, the water loss directly affects the embryo, with the exposed radicle being the most rapidly affected. The thick cotyledons enclose the small plumule and it dehydrates last. In species with minute, rudimentary embryos (*O. novogranatensis*, *M. guianensis*), the sequence of dehydration is seedcoat ¢ peripheral endosperm and radicle ¢ remaining embryo ¢ inner endosperm.

Seed viability deteriorates rapidly upon embryo dehydration. *Otoba novogranatensis* is more sensitive to desiccation than *Minquartia guianensis,* due to endosperm rumination and tegmen vascularization. *Caryocar costaricense* has a curved, accumbent embryo and the plumule and radicle dehydrate immediately after endocarp dehydration. The dehydration of *Lecythis ampla* seeds first affects the seedcoat and then the meristematic poles (radical and apical); they die instantly after seedcoat dehydration.

The seeds enclosed in berries (*C. brasiliense*) or drupes (*C. costaricense, M. guianensis*) are protected by the pericarp tissues, which help maintain seed moisture. In these cases, the functional unit is the fruit and dehydration is slower. The moisture level below which a seed loses its viability varies from one seed to another. Variations are found among seeds collected from the same tree as well as from different trees, zones, seasons, or years. The recalcitrant behavior seems to be genetically determined and its genetic base is still not well understood. The variations found can be explained if the seed history from flower inception to seed dispersal and germination is analyzed carefully.

The difference found between temperate and tropical recalcitrant seeds must be added to the gradient found in recalcitrance manifestation. Bonner and Vozzo established two categories: temperate recalcitrant seeds and tropical recalcitrant seeds. The first cannot be dried but can be stored for 3 to 5 years at near freezing temperatures; the latter cannot be dried and are killed by temperatures below 10 to 15 °C, depending on the species.

Unfortunately, very limited information exists on the tissular, cellular, and biochemical alterations produced by dehydration in the recalcitrant seeds, as well as on the appropriate strategies and mechanisms to manage them under storage conditions. The diversity of types found in recalcitrant seeds is preliminary and further complicates the issue. The problem of cellular desiccation is complex; it seems to involve genetic components that lead to mechanisms of cellular protection. These mechanisms limit the cell damage produced by seed dehydration and promote cellular repair, reversing the changes induced by water loss. The accumulation of protecting substances in the tolerant tissues is quite possible. The dehydrines, disaccharides (sacarose), and oligosaccharides (raffinose and stachyose) may have an important role in the stabilization and maintenance of the membrane system and other sensitive systems.

Bibliography

Black, Michael, Bradford, K. J., Vazquez-Ramos, Jorge (ed.). *Seed Biology: Advances and Applications: Proceedings of the Sixth International Workshop on Seeds*, Mérida, México, 1999.

Bradbeer, J. W., *Seed Dormancy and Germination*, New York: Chapman and Hall, 1988.

Deno, N. C., *Seed Germination: Theory and Practice*, State College, PA : Norman C. Deno, 1993.

Ellis, R. H. (ed.), *Basic and Applied Aspects of Seed Biology: Proceedings of the Fifth International Workshop on Seeds*, Reading, 1995. Boston: Kluwer Academic Publishers, 1997

______________. (ed.), *Long-term Seed Storage of Major Temperate Fruits*, Rome: IBPGR Secretariat, 1985.

Holliday, Paul, *A Dictionary of Plant Pathology*, New York: Cambridge University Press, 1998.

Jensen, Hans Arne, *Bibliography on Seed Morphology*, Brookfield, VT: A.A. Balkema, 1998.

Mathur, S. B., Jørgensen, Johs (ed.), *Seed Pathology*, Wageningen, Netherlands: Technical Centre for Agricultural and Rural Co-operation, 1992. Proceedings of the CTA Seminar Held at Copenhagen, Denmark, on 20-25 June 1988.

Multon, J. L. (ed.), *Preservation and Storage of Grains, Seeds, and Their By-products: Cereals, Oilseeds, Pulses, and Animal Feed*, New York, N.Y.: Lavoisier Pub., 1988.

Obroucheva, Natalie V., *Seed Germination: A Guide to the Early Stages*, Leiden: Backhuys Publishers, 1999.

Sen, David N., Sher Mohammed, (ed.), *Marvels of Seed: Proceedings of International Symposium on* Environmental Influences on Seed and Germination Mechanism - Recent Advances in Research & *Technology,* Jodhpur, India: Dept. of Botany, University of Jodhpur, 1991.

Taylor, Alan G., Huang, Xue-lin, (ed.), *Progress in Seed Research: Conference Proceedings of the Second International Conference on Seed Science and Technology (2nd ICSST),* May 12-16, 1997, Geneva, N.Y.: Communication Services, NYS Agricultural Experiment Station, 1997.

Werker, Ella, *Seed Anatomy,* Berlin: G. Borntraeger, 1997.

Index